*Intelligent Vibration Control
in Civil Engineering Structures*

Intelligent Vibration Control in Civil Engineering Structures

Zhao-Dong Xu
Southeast University, Nanjing, Jiansu, China

Ying-Qing Guo
Nanjing Forestry University, Nanjing, Jiansu, China

Jun-Tao Zhu
Southeast University, Nanjing, Jiansu, China

Fei-Hong Xu
Southeast University, Nanjing, Jiansu, China

AMSTERDAM • BOSTON • HEIDELBERG • LONDON
NEW YORK • OXFORD • PARIS • SAN DIEGO
SAN FRANCISCO • SINGAPORE • SYDNEY • TOKYO
Academic Press is an imprint of Elsevier

Academic Press is an imprint of Elsevier
125 London Wall, London EC2Y 5AS, United Kingdom
525 B Street, Suite 1800, San Diego, CA 92101-4495, United States
50 Hampshire Street, 5th Floor, Cambridge, MA 02139, United States
The Boulevard, Langford Lane, Kidlington, Oxford OX5 1GB, United Kingdom

Copyright © 2017 Zhejiang University Press Co., Ltd. Published by Elsevier Inc. All rights reserved.

No part of this publication may be reproduced or transmitted in any form or by any means, electronic or mechanical, including photocopying, recording, or any information storage and retrieval system, without permission in writing from the publisher. Details on how to seek permission, further information about the Publisher's permissions policies and our arrangements with organizations such as the Copyright Clearance Center and the Copyright Licensing Agency, can be found at our website: www.elsevier.com/permissions.

This book and the individual contributions contained in it are protected under copyright by the Publisher (other than as may be noted herein).

Notices
Knowledge and best practice in this field are constantly changing. As new research and experience broaden our understanding, changes in research methods, professional practices, or medical treatment may become necessary.

Practitioners and researchers must always rely on their own experience and knowledge in evaluating and using any information, methods, compounds, or experiments described herein. In using such information or methods they should be mindful of their own safety and the safety of others, including parties for whom they have a professional responsibility.

To the fullest extent of the law, neither the Publisher nor the authors, contributors, or editors, assume any liability for any injury and/or damage to persons or property as a matter of products liability, negligence or otherwise, or from any use or operation of any methods, products, instructions, or ideas contained in the material herein.

British Library Cataloguing-in-Publication Data
A catalogue record for this book is available from the British Library

Library of Congress Cataloging-in-Publication Data
A catalog record for this book is available from the Library of Congress

ISBN: 978-0-12-405874-3

For Information on all Academic Press publications
visit our website at https://www.elsevier.com

Publisher: Jonathan Simpson
Acquisition Editor: Glyn Jones
Editorial Project Manager: Naomi Robertson
Production Project Manager: Susan Li
Designer: Mark Rogers

Typeset by MPS Limited, Chennai, India

Contents

Preface .... *xi*

Chapter 1: Introduction 1
   1.1 Earthquake and Wind Disasters 1
      1.1.1 Earthquake Disaster 1
      1.1.2 Wind Disaster 6
   1.2 Structure Vibration Control 8
      1.2.1 Basic Principles 8
      1.2.2 Classification 9
      1.2.3 Structure Intelligent Control 16

Chapter 2: Intelligent Control Strategies 21
   2.1 Equations of Motion of Intelligent Control System 21
   2.2 Classical Linear Optimal Control Algorithm 22
      2.2.1 LQR Optimal Control 23
      2.2.2 LQG Optimal Control 25
   2.3 Pole Assignment Method 27
      2.3.1 Pole Assignment Method With State Feedback 28
      2.3.2 Pole Assignment Method With Output Feedback 29
   2.4 Instantaneous Optimal Control Algorithm 30
   2.5 Independent Mode Space Control 33
      2.5.1 Modal Control Based on State Space 33
      2.5.2 Modal Control Based on Equation of Motion 34
   2.6 H_∞ Feedback Control 35
      2.6.1 H_∞ Norm 36
      2.6.2 H_∞ Feedback Control 37
   2.7 Sliding Mode Control 38
      2.7.1 Design of Sliding Surface 39
      2.7.2 Design of Controller 40

2.8 Optimal Polynomial Control 41
    2.8.1 Basic Principle 42
    2.8.2 Applications 45
2.9 Fuzzy Control 45
    2.9.1 Basic Principle 46
    2.9.2 Design of Fuzzy Controller 46
2.10 Neural Network Control 50
    2.10.1 Basic Principle 50
    2.10.2 Learning Method 51
2.11 Particle Swarm Optimization Control 52
    2.11.1 Basic Principle 53
    2.11.2 Design Procedure of the PSO Algorithm 55
2.12 Genetic Algorithm 56
    2.12.1 Basic Principle 56
    2.12.2 Procedure of GA 57
    2.12.3 GA Control Realization 60

Chapter 3: Active Intelligent Control 63

3.1 Principles and Classification 63
    3.1.1 Buildup of Systems 63
    3.1.2 Basic Principles 64
    3.1.3 Classification 64
3.2 Active Mass Control System 65
    3.2.1 Basic Principles 65
    3.2.2 Construction and Design 67
    3.2.3 Mathematical Models and Structural Analysis 69
    3.2.4 Experiment and Engineering Example 72
3.3 Active Tendon System 74
    3.3.1 Basic Principles 74
    3.3.2 Construction and Design 75
    3.3.3 Experiment and Engineering Example 77
3.4 Other Active Control System 79
    3.4.1 Form and Principles 79
    3.4.2 Analysis and Tests 81

Chapter 4: Semiactive Intelligent Control 85

4.1 Principles and Classification 85
    4.1.1 Basic Principles 85
    4.1.2 Classification 86

4.2	MR Dampers		87
	4.2.1	Basic Principles	87
	4.2.2	Construction and Design	89
	4.2.3	Mathematical Models	94
	4.2.4	Analysis and Design Methods	99
	4.2.5	Tests and Engineering Applications	101
4.3	ER Dampers		110
	4.3.1	Basic Principles	110
	4.3.2	Construction and Design	112
	4.3.3	Mathematical Models	113
	4.3.4	Analysis and Design Methods	115
	4.3.5	Tests and Engineering Applications	115
4.4	Piezoelectricity Friction Dampers		117
	4.4.1	Basic Principles	118
	4.4.2	Construction and Design	118
	4.4.3	Mathematical Models	119
	4.4.4	Analysis and Design Methods	123
	4.4.5	Tests and Engineering Applications	124
4.5	Semiactive Varied Stiffness Damper		126
	4.5.1	Basic Principles	126
	4.5.2	Construction and Design	127
	4.5.3	Mathematical Models	128
	4.5.4	Analysis and Design Methods	129
	4.5.5	Tests and Engineering Applications	131
4.6	Semiactive Varied Damping Damper		132
	4.6.1	Basic Principles	133
	4.6.2	Construction and Design	133
	4.6.3	Mathematical Model	134
	4.6.4	Analysis and Design Methods	136
	4.6.5	Tests and Engineering Applications	137
4.7	MRE Device		138
	4.7.1	Basic Principles	138
	4.7.2	Construction and Design	140
	4.7.3	Mathematical Models	142
	4.7.4	Analysis and Design Methods	144
	4.7.5	Tests and Engineering Applications	145

Chapter 5: Design and Parameters Optimization on Intelligent Control Devices 151

5.1	Design and Parameters Optimization on MR Damper		151
	5.1.1	Design on MR Damper	151
	5.1.2	Parameters Optimization on MR Damper	156

5.2 Design and Parameters Optimization of MRE Device 160
    5.2.1 Parameter Optimization for Magnetic Circuit 160
    5.2.2 Magnetic Circuit FEM Simulation.. 165
5.3 Design and Parameters Optimization on Active Control... 166
    5.3.1 Design and Parameters Optimization Based on Feedback Gain 167
    5.3.2 Design and Parameters Optimization Based on Minimum Energy Principle... 169
    5.3.3 Design and Parameters Optimization Based on Fail-Safe Reliability 171

Chapter 6: Design and Study on Intelligent Controller *173*

6.1 Design of Intelligent Controller.. 173
    6.1.1 The Design of the Acceleration Responses Collection............................. 173
    6.1.2 The Design of the Microcontroller.. 175
6.2 Experimental Study on Intelligent Controller 177

Chapter 7: Dynamic Response Analysis of the Intelligent Control Structure *181*

7.1 Elastic Analysis.. 181
    7.1.1 Mathematical Model of Structures... 181
    7.1.2 Determination of the Control Force of the MR Damper 185
    7.1.3 Numerical Analysis 186
7.2 Elasto-Plastic Analysis Method 187
    7.2.1 Restoring Force Model 187
    7.2.2 Processing of Turning Points 188
    7.2.3 Elasto-Plastic Stiffness Matrix 190
7.3 Dynamic Response Analysis by SIMULIK... 192
    7.3.1 Simulation of the Controlled Structure 192
    7.3.2 Numerical Analysis 193

Chapter 8: Example and Program Analysis.. *201*

8.1 Dynamic Analysis on Frame Structure With MR Dampers 201
    8.1.1 Structural and Damper Parameters... 201
    8.1.2 Semiactive Control Strategy 201
    8.1.3 Results and Analysis... 204
8.2 Dynamic Analysis on Long-Span Structure With MR Dampers.................................. 208
    8.2.1 Parameters and Modeling 208
    8.2.2 Wind Load Simulation... 209
    8.2.3 Semiactive Control Strategy 211
    8.2.4 Results and Analysis... 213

8.3 Dynamic Analysis on Platform With MRE Devices 215
    8.3.1 Modeling and Parameters 217
    8.3.2 Semiactive Control Strategy 220
    8.3.3 Results and Analysis.. 222
8.4 SIMULINK Analysis Example 223
    8.4.1 The SIMULINK Example of the Structure Without Dampers................ 224
    8.4.2 The SIMULINK Example of the Controlled Structure............................ 226
8.5 Particle Swarm Optimization Control Example 239
    8.5.1 Structural and Damper Parameters... 239
    8.5.2 The PSO Optimization Control 240
    8.5.3 Results and Analysis.. 241
8.6 Active Control Example.. 243
    8.6.1 Modeling and Parameters 243
    8.6.2 Active Control Strategy 246
    8.6.3 Results and Analysis.. 247

References **249**
Index **259**

Preface

Structure vibration control is recognized as an effective method and possesses a promising prospect in reducing dynamic responses of structures by using additional vibration mitigation devices. It can be divided into two main categories: the passive control and the intelligent control. Passive control is a technology without external energy input. Some changes in structure dynamic property can be made by mounting shock absorptions or energy isolation devices, which dissipate, divert, or block the transmission of vibration energy. The control devices cannot change characteristic factors, therefore the passive control has many merits, such as simple construction, low cost, easy to maintain, not requiring external energy. Some passive control devices are quite mature and widely used in practical engineering, are even organized into the related codes. Passive control is mainly divided into two categories: vibration isolation and vibration mitigation. Intelligent control includes active control and semiactive control. Active control is a technology that needs for external energy, aiming at protecting structures or equipment. According to the modern control theory, it observes, tracks, and predicts the input of ground motion and structural responses on line, then analyzes the calculated results and imposes the opposite control force by a servomechanism to achieve selfregulating structure responses under dynamic loads. Semiactive control system has a great practical value and good engineering application, which combines the reliability of passive control and flexibility of active control. It can supersede the active control with improvements in some control laws and has a simple construction and stable performance. Thus, the semiactive control system can get over the constrained energy delivery in active control and narrow-band range in passive control. As a hot field, structural vibration control has developed rapidly and research achievements have been widely applied to engineering practice.

Intelligent vibration control has attracted a great deal of attention due to the precise vibration mitigation effect and the adaptability of control devices for different excitations. With particular emphasis on the alleviation of the dynamic responses of complex large-scale or important civil structures, intelligent vibration control will be implemented in accordance with different dynamic excitations. Obviously, an intelligent vibration control system consists of sensor system, control system, control device, and other components, therefore intelligent vibration control involves in some research topics including control devices, intelligent control strategies, intelligent controllers, and intelligent controlled structures. Not as mature as passive vibration control, many challenging topics of intelligent vibration control are still required to be studied further. At the same time, intelligent vibration control in civil engineering structures incorporates interdisciplinary technologies, including civil engineering, automation, mechanics, mechanical engineering, electronic engineering, etc. Therefore systematic books about intelligent vibration control are relatively rare. In view of this, the authors have written this book which provides a working knowledge of this exciting and fast expanding field and brings up-to-date current research and world-wide development of intelligent vibration control, and hope to help the relative researchers, teachers, and students in design, analysis, and application of intelligent vibration control.

The book is divided into eight chapters. In Chapter 1, the main dynamic excitations of civil engineering structures which refers to earthquake and wind excitations are introduced firstly, and then structural vibration control, especially for structural intelligent control, is classified and described. In Chapter 2, the equations of motion of intelligent control system are erected firstly, and then different intelligent control strategies for intelligent vibration control in civil engineering structures are introduced. In Chapter 3, active intelligent controls including principles and classification, active mass control system, active tension control system, and other active control systems are introduced. In Chapter 4, semiactive intelligent controls including principles and classification, magnetorheological damper, electrorheological damper, piezoelectricity friction damper, semiactive varied stiffness damper, semiactive varied damping damper, and magnetorheological elastomer device are introduced. In Chapter 5, taking magnetorheological damper, magnetorheological elastomer device, and active mass damper as examples, design and parameter optimization on intelligent control devices are presented. In Chapter 6, the intelligent controller is designed and experimented. In Chapter 7, dynamic response analysis methods of the intelligent control structure, including elastic, Elasto-plastic, and SIMULINK analysis method, are described. In Chapter 8, some intelligent control examples are introduced systematically, and the corresponding programs are compiled.

It is a great pleasure to acknowledge the contributions to the writing of this book by the researchers in my group, Xiang-Cheng Zhang, Jun Dai, Pan-Pan Gai, Ye-Shou Xu, Yang Yang, Si Suo, Yu-Liang Zhao, Meng Xu, Cheng Wang, Teng Ge, Qian-Wei Jiang, Chao Xu, Waseem Sarwar, Yan-Long Su, Da-Huan Jia, Ling-Feng Sha, Bing-Bing Chen, Tao Liu, Qian-Qiu Yang, Xing-Huai Huang, Jun-Jian Wang, and An-Nan Miao. Special thanks should be given to Prof. Jin-Ping Ou, Prof. Fu-Lin Zhou, Prof. B.F. Spencer, Prof. Fuh-Gwo Yuan, Prof. Ya-Peng Shen, Prof. Hong-Tie Zhao, Prof. Yong Lu, and Ms. Xiu-Fang Wu for their help and encourage. Financial supports for the research of our group described in this book are provided by National Science Fund for Distinguished Young Scholars, National High Technology Research and Development Program (863) with grants number

2009AA03Z106, National Natural Science Funds of China with granted numbers 11176008, 61004064, 11572088, Jiangsu Province Brace Program with granted number BE2010069, Jiangsu Natural Science Funds with grants numbers BK20140025, BK2005410, BK20141086, BE2015158. These supports are gratefully acknowledged.

Due to that the contents of the book are much involved in cutting-edge research technology, some errors maybe occur in the book. The authors appreciate any careful comment of experts and readers sincerely.

Corresponding author

Nanjing, China

Jan. 8, 2016

CHAPTER 1

Introduction

In the design of many civil engineering structures, the primary static loads, including dead and live loads, are not adequate. Instead, the structures must be analyzed and designed by considering dynamic excitations, such as winds, earthquakes, or waves. Dynamic loads are usually the most important factors causing damage or collapse of structures. The time-varying and inertial characteristics of dynamic problem make it more complex than the static problem. Earthquake and strong wind are the main dynamic excitations leading to the disasters of civil engineering structures.

1.1 Earthquake and Wind Disasters

1.1.1 Earthquake Disaster

Earthquake, as one of the main dynamic excitations, which has potential disastrous consequences to civil engineering structures, must be carefully considered. The occurrence of earthquakes has characteristics that are random, unexpected, and uncertain. In accordance with the reasons of the formation, earthquakes can be divided into tectonic earthquakes, volcanic earthquakes, collapse earthquakes, and induced earthquakes. Throughout the world, the majority of earthquakes are tectonic earthquakes, which are caused by sudden fracturation due to crustal movement, as shown in Fig. 1.1. The failure position of rock formations is called the hypocenter, and the projection of the earthquake origin to the ground is called the epicenter. The depth between hypocenter and the epicenter is called focal depth. According to the focal depth, earthquakes can be divided into shallow-focus earthquakes (focal depth is less than 70 km), intermediate-focus earthquakes (focal depth is between 70 and 300 km) and deep-focus earthquakes (focal depth exceeds 300 km).

There are three major seismic belts in the world, as shown in Fig. 1.2. The Circum-Pacific seismic belt, which accounts for 80% of the world's shallow-focus earthquakes, 90% of intermediate-focus earthquakes, and where almost all deep-focus earthquakes have occurred, are located in the Pacific Rim, including New Zealand, Indonesia, Philippines, Chinese Taiwan, Japan, Aleutian Islands, and the west coasts of North America and South America. The Eurasian seismic belt contains the Himalayas, Burma, India, Pakistan, Iran, Turkey, and the Mediterranean region. Ocean ridge seismic belt, where the seismic occurrence frequency is not high, includes the Pacific, Atlantic, Indian, and mid-ocean

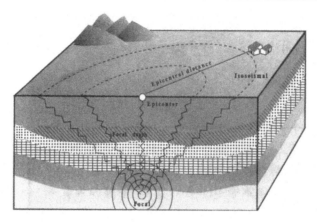

Figure 1.1
Schematic diagram of seismic structure.

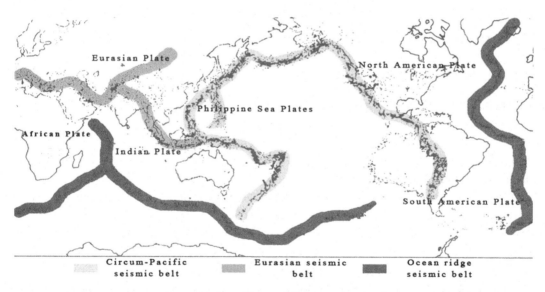

Figure 1.2
Seismic belts around the world.

ridges of the Arctic Sea. In the 20th century, there have been more than 2600 devastating earthquakes in the world, which has led to about 1,260,000 deaths, nearly a million people disabled, and hundreds of billion dollars worth of property damage and the like. At the start of the 21st century, statistics show that the level of seismic activity was significantly strengthened, and the strength of earthquakes also increased. All of these present a challenge to civil engineering researchers and engineers.

Vibrations caused by an earthquake propagate from the hypocenter to all directions and release energy in the form of seismic waves. There are two kinds of seismic waves: one consists of body waves, including P-wave and S-wave, and the other are surface waves, including Rayleigh wave and Love wave. The energy and strength of earthquakes are evaluated by earthquake magnitude, with only one magnitude for each earthquake. The Richter magnitude, proposed by Charles Richter in 1935, is used to assess the strength of earthquakes internationally. However, earthquake magnitude cannot reflect the degree of destruction in an area. Usually, seismic intensity is adopted to reflect the destroy effect of an earthquake in a certain area, which is given more attention by designers or engineers. Each nation has drawn up the intensity scale according to their circumstances. The Modified Mercalli Intensity scale, which has 12 identified standards from I (no feeling) to XII (complete destruction), is used in the USA. The seismic intensity in China also ranges from I (no feeling) to XII (complete destruction).

Earthquake ground motion with strong randomness is very complex; its characteristics are usually described by three basic elements, namely, the excitation amplitude, the frequency spectrum, and the duration time. The excitation amplitudes usually refer to the peak values of ground motion acceleration, velocity, or displacement. The greater the amplitudes of excitations are, the more severe the damage to civil structures. The frequency spectrum shows the frequency-domain distribution characteristics of seismic waves, and it can be represented by a response spectrum, a power spectrum, or a Fourier spectrum. Earthquakes with different spectral characteristics have different effects on civil engineering structures with different dynamic characteristics. The duration time is the duration of an earthquake, and the longer the duration time is, the more severe the damage to civil structures may be.

It is well known that the majority of lost lives and destroyed properties during an earthquake are caused by the destruction of civil engineering structures. For instance, on March 11, 2011 in Japan, an earthquake with a 9.0 magnitude resulted in a hydrogen coolant explosion of one of the Fukushima Daiichi Nuclear Power Station units, followed by a nuclear leak of the Units 3, 2, and 4, which triggered a nuclear crisis in East Asia. Similarly, a lot of civil structures were destroyed in the Tangshan earthquake on July 28, 1976 as well as in the Wenchuan earthquake on May 12, 2008, as shown in Figs. 1.3 and 1.4.

In order to reduce the number of disasters of civil engineering structures due to earthquakes, it is necessary to apply reliable anti-earthquake protection measures on structures. This requires seismic actions and seismic responses of structures that can be calculated relatively accurately. Seismic action, which is not like the direct-acting style of other dynamic excitations, is applied to structures through supporting the movement indirectly. For civil engineering structures, how to calculate seismic responses, including

Figure 1.3
Damage in Tangshan earthquake.

Figure 1.4
Damage in Wenchuan earthquake.

displacement, acceleration, velocity, and strain responses? Currently, seismic analysis and design of civil engineering structures can be divided into static [1], response spectrum [2,3], and dynamic analysis methods [4]. Dynamic analysis methods include the time-domain method and the frequency-domain method.

Static analysis method is a simplification for seismic analysis of structures. This method assumes that structures are based on a rigid foundation, and the inertia force under the ground motion acceleration is regarded as a static force acting on structures. The formula of the inertia force can be expressed as

$$F = kG \tag{1.1}$$

where k is the seismic coefficient. The seismic coefficient can be specified according to different areas. It is a significant improvement to propose the static analysis method for the design of engineering structures. However, the seismic action F has nothing to do with the dynamic characteristics of engineering structures. In fact, the static analysis method can only be established under conditions where the natural period of the structure is far smaller than the predominant period of the site.

The response spectrum method has taken dynamic characteristics of structures and seismic ground motions into account. The seismic action can be calculated in accordance with

$$F = k\beta(T)G = \alpha G \tag{1.2}$$

where α is the design spectrum. Considering the randomness of the seismic ground motions, the design spectrum can be obtained through the representative average response spectrum under the same kinds of sites for different ground-motion-acceleration excitations. The response spectrum method, which is widely used around the world, considers not only the affection of ground motions and the soil properties of the site, but also the dynamic characteristics of structures, such as the natural period and damping ratio. However, since the response spectrum reflects the relationship between the maximum action and the natural period of structures under given seismic motion for a single degree of freedom (SDOF) elastic systems, this method cannot present real dynamic responses of structures during earthquakes.

The time history analysis method is first to divide the whole seismic wave into many time steps, and then applying numerical integration for the vibration differential equation of structures during each step, finally to get dynamic responses of the entire time-domain process. The time history analysis method has incomparable advantages over the response spectrum method. Firstly, the method considers the characteristic of the duration time. Multiple seismic records have proved that the earthquake damage was significantly associated with the duration time of earthquakes. For example, on July 9, 1971, in San Fernando, California, an earthquake with a magnitude of 6.6 caused a larger loss because of the long duration time. Secondly, the time history analysis method can get responses of structures during the elastoplastic stage, which also takes the material and geometric nonlinearities into account.

1.1.2 Wind Disaster

Wind is another main dynamic excitation for civil engineering structures. For a long time people have made use of wind loads for their needs, and yet many people have suffered from tremendous wind disasters. Wind disasters cause significant losses of life and property every year, and is listed within the top three of the leading causes of natural disasters. Therefore, a concern for wind load is one of the main loads in the design of civil engineering structures, such as large-span structures, high-rise structures, and guyed mast structures.

Wind is a massive air flow phenomenon. The inner air pressure in different parts is changes with air density, terrain, temperature, latitude, and longitude. Therefore, differential pressure exists in neighboring regions, and the pressure difference and the earth's rotation make the air flow resulting in the formation of wind. According to the reasons for the formation, the wind can be divided into a number of categories: monsoon, dry-hot wind, tropical cyclone, water elect, valley wind, foehn storm, and heat island circulation, etc. The main wind disasters to engineering structures are caused by tropical cyclones. In accordance with the strength, wind can be divided into 0 to 12 levels, which is suggested by the world meteorological organization presently.

Wind velocity is associated with landform sites. Due to the friction, the wind velocity has a tendency to be reduced near the surface of the site. The wind velocity is not affected by the friction only when the height exceeds a value from the ground where the wind speed is called a gradient velocity and the height is called the height of gradient wind. The air layer near the ground and below the height of gradient wind is the so-called friction layer, where the wind is affected by geographical position, topographic conditions, ground roughness, height, temperature variation, and other factors. The wind-resistant design of civil engineering structures should consider the wind characteristics, including the law of wind speed variation, horizontal angle of wind speed, the strength of fluctuating wind speed, periodic component, and spatial correlation of wind speed, etc.

Generally speaking, wind speed time curve includes the mean wind and the fluctuating wind. The mean wind, which is also called stable wind, is mainly affected by the long-period component of wind and its features include the average wind direction, the average wind velocity, wind speed profile, and wind frequency curves. Since the period of mean wind is much longer than the natural vibration period of civil structures, the mean wind can be regarded as a static force. Fluctuating wind is also called gust pulsation, which is the short-period component of the wind with only a few seconds generally. Fluctuating wind's features include the fluctuating wind speed, the fluctuation coefficient, changes in wind direction, strength of turbulence, turbulence integral scale, fluctuating wind power spectrum, and coefficient of spatial correlation, etc. Since the strength of fluctuating wind varies with time and its period is close to the vibration period of some engineering structures, responses due to the fluctuating wind are usually calculated by dynamic or quasi-static methods.

In addition to earthquake disasters, wind disasters are also a primary concern for civil engineering structures because they can also lead to large population casualties and economic losses. Strong wind disasters, which are generally up to VI level on average but can go up to level VIII level for disasters of instantaneous destruction, are regarded as one of the top three cause of natural disasters. Almost all countries and regions located in the west coast are affected by tropic ocean cyclones. And China is one of the countries suffered most severely from typhoons. The number of typhoons and tropical storm landings in China can be up to seven times per year, according to statistical data from 1971 to 2000. For civil engineering structures, strong winds mainly cause the failure of engineering structures in the form of cracking, damage, and collapse, especially for high, thin, and long flexible engineering structures. The reliable wind vibration response analysis and wind-resistant design are the key steps to ensure the safety of structures.

Wind vibration response analysis includes simulation of wind loads and response calculation. As with the earthquake response analysis described earlier, wind vibration response analysis methods are divided into two kinds of methods: dynamic and quasi-static methods. Dynamic simulations of wind loads generally include *harmonic superposition method* [5], *linear filtering method*, the *corrected Fourier spectroscopy method* [6], etc. The structural wind vibration responses can be then obtained by a time-domain method or a frequency-domain method similar to seismic responses analysis. Dynamic analysis methods are precise but complex for designers. Therefore, quasi-static methods are adopted in most of codes in the world. For example, for the Chinese wind code, wind pressure applied to structures is written as

$$\omega_k = \beta_z \mu_z \mu_s \omega_0 \tag{1.3}$$

where the coefficient μ_z is a reflection of the pressure/velocity distribution with the height of the structures. The coefficient μ_z varies with the ground roughness category and terrain conditions. The shape coefficient of wind load μ_s is the ratio between the actual pressure (suction) caused by wind and the pressure (suction) calculated by a formula for the relation between wind speed and wind pressure. The coefficient β_z is the ratio between the total wind loads and the average wind loads, which considers the fluctuating wind load simply by the code. The basic wind pressure ω_0 can be obtained by statistical methods according to related specifications. After the wind pressure is determined, the displacement and inner forces can be calculated by *the D-value method* [7].

Wind resistance design of civil engineering structures aims to ensure that civil engineering structures can bear the maximum wind loads and withstand the dynamic action during the construction phase and the entire life cycle. Because natural winds can lead to different kinds of structural vibrations, the critical velocity of the flutter and buffeting must have enough safety compared to the design velocity of the structures to ensure the structural wind resistance stability during different stages. It is also a requirement to control the maximum amplitudes induced by vortex vibration and buffeting within an acceptable range to avoid structural fatigue and ensure structural comfort. Engineers should control structures' wind-resistance stability and reduce wind-induced vibration amplitude by modifying the design or using aerodynamic measures, structural measures, and other control measures when the original design cannot meet the requirements of the wind resistance.

Given the serious consequences of disasters caused by wind, importance has been attached to wind engineering research internationally, and all countries in the world have made *codes* for wind resistance of civil engineering structures. The accurate simulation of wind loads and the reliable wind responses analysis govern the wind resistant design of structures directly. In wind resistant design, the quasi-static methods can be accurate enough for ordinary structures. However, for the design of special or important structures, such as long-span structures and high-rise buildings, wind resistance design should be accomplished based on dynamic methods.

1.2 Structure Vibration Control

Structure vibration control technology is an emerging technology for reducing the vibration of civil engineering structures induced by earthquakes or wind, which is different from traditional methods that rely on structural strength and stiffness to reduce structural vibration.

1.2.1 Basic Principles

The vibration control mechanism can be described as follows: when adding appropriate vibration isolation or mitigation devices in structures, vibration will be isolated or mitigated, so that the main structures are protected from damage by these devices.

The equation of motion of the structural system under dynamic loads can be expressed as:

$$[M]\{\ddot{x}(t)\} + [C]\{\dot{x}(t)\} + [K]\{x(t)\} = \{P(t)\} + \{F_D(t)\} \tag{1.4}$$

where $[M]$, $[C]$, $[K]$ are the structural mass, damping, and stiffness matrices, respectively; $\{\ddot{x}(t)\}$, $\{\dot{x}(t)\}$, $\{x(t)\}$ are the acceleration, velocity, and displacement response vectors, respectively; $\{P(t)\}$ is the external excitation load vector; and $\{F_D(t)\}$ is the controlled force vector provided by energy dissipation devices. For vibration mitigation by using energy dissipation devices, $\{F_D(t)\}$ is usually in the reverse direction with $\{P(t)\}$, especially for the intelligent control described in this book.

The basic principles of structural vibration control can be interpreted from the energy relationship of the structure at any moment during vibration [8]:

$$E_{in} = E_p + E_k + E_h + E_d \tag{1.5}$$

where E_{in} is the energy input into the structure from the external excitation, such as earthquake or wind, E_p is the potential energy of the structural system, E_k is the kinetic energy of the structural system, E_h is the energy dissipated by the structural system through inelastic or other forms of action including structural damping, and E_d is the energy dissipated by added energy dissipation devices. E_h is usually contributed by energy dissipation due to the structural damping and structural elastoplastic deformation. The structural elastoplastic deformation will result in structural damage or even destruction. Therefore, it is expected to protect the main structure by reducing E_h. The technique of vibration isolation decreases the input energy E_{in} through adding isolation devices between the base and the upper structure. This will lead to a decrease of the energy dissipation due to structural elastoplastic deformation and reduce the dynamic responses of the main structure. For the technique of vibration mitigation, although there is no decrease for the input energy E_{in}, there is the additional energy dissipation E_d contributed by added energy dissipation devices. This will also lead to the decrease of the energy dissipation due to structural elastoplastic deformation and reduce the dynamic responses of the main structure.

1.2.2 Classification

Structure vibration control systems can be divided into passive control systems and intelligent control systems in accordance with whether the system needs an external energy or not, as illustrated in Fig. 1.5. This section will briefly describe the passive control systems and Section 1.2.3 will briefly introduce the intelligent control system.

The passive control system does not require external energy and depends on the responses of the structure and external excitation information. Here the control devices cannot change characteristic factors; therefore, the passive control system has many merits, such as simple construction, low cost, easy to maintain, and not requiring external energy. Some passive control devices are quite mature and widely used in practical engineering and they even are organized into the relative codes. The passive control system is mainly divided into two categories: vibration isolation and vibration mitigation.

1.2.2.1 Vibration isolation

Vibration isolation system is that setting of isolation devices between the upper structure and the foundation, as shown in Fig. 1.6, and this will reduce the transmission of vibration energy. The methods of vibration isolation usually include rubber isolation, sliding isolation, hybrid isolation, etc.

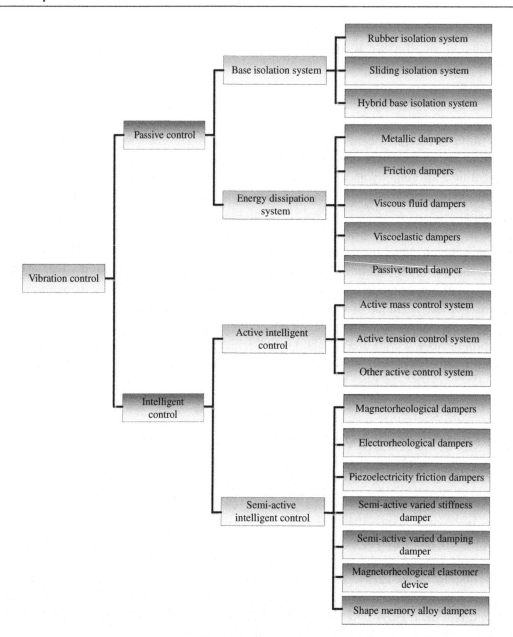

Figure 1.5
Classification of vibration control.

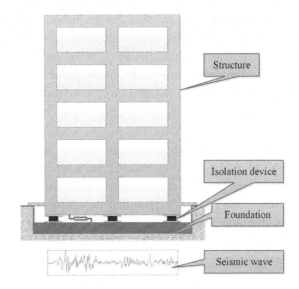

Figure 1.6
Structure vibration isolation.

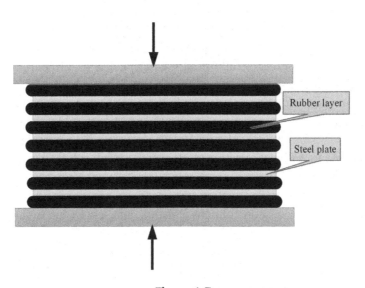

Figure 1.7
Rubber isolation bearing.

1. Rubber isolation

   Rubber bearing, also known as sandwich (or laminated/multi) rubber isolation pad, consists of rubber and steel plane alternately. Rubber bearings are the most mature and widely used isolation devices, as shown in Fig. 1.7. In order to meet the damping requirements of the isolation system, lead plug or high damping rubber is adopted in rubber bearings. The rubber bearing is an ideal device not only having the advantages of strong vertical bearing capacity, small horizontal stiffness and large horizontal allowable displacement, but also withstanding the vertical earthquake excitations with certain strength. However, the device has poor tension properties and the effect of vertical isolation is not obvious. In the recent years, some multi-dimensional earthquake isolation devices are proposed based on the ordinary rubber bearing [9,10].

2. Sliding isolation

Sliding isolation relies on sliding support elements installed between the basement and the upper structure. Sliding support elements can reduce transmission of vibration energy effectively due to energy consumption by relative sliding motion and friction. Theoretical and practical studies [11,12] show that the sliding isolation system has excellent capacity for vibration isolation. The main advantage of the sliding isolation system is that its isolation effect is affected by the frequency of slight ground motion, and resonance phenomenon does not occur easily. At the same time, the construction of the sliding isolation system is simple and the cost is lower. The sliding isolation system has some drawbacks, such as bad restoring capacity due to no lateral stiffness and an instability of the friction coefficient.

3. Hybrid isolation

The hybrid isolation system is constituted by two or more isolation systems in a certain way (series, parallel, and series-parallel, etc.) for better vibration isolation effect. A hybrid isolation system makes full use of the advantages of different isolation systems and overcomes shortcomings of different components. For example, when the sliding isolation system and the rubber isolation system are connected in parallel, the former has better load-carrying capacity and can decrease the number of rubber bearings, and the latter can provide restoring force. Hybrid isolation systems have complex hysteresis characteristics due to different isolation components, which can lead to difficult analysis.

4. Other isolation systems

Besides the above three main isolation systems, some new forms of isolation systems, such as inter-story isolation, short columns isolation, and suspension isolation appear under a special application environment [13,14]. An inter-story isolation system is to set the isolation layer between the pillars and floors of the structures for reducing the structural responses. Compared to the traditional isolation system, the only difference is the position of the isolation layer. The short columns isolation system is to set short columns in the corresponding parts of the structure, which takes advantage of the elastoplastic energy consumption of steel tube concrete columns and the friction energy consumption between the upper and lower bearings. Suspension isolation structures are realized by linking the structures and supports with flexible slings or hanger rods, and this system works as a gravity pendulum. The suspension isolation system reduces structural responses in all directions. With the development of isolation technology, new seismic isolation devices and systems will be proposed in the future.

1.2.2.2 Vibration mitigation

A vibration mitigation system concerns installing energy-consuming device in the structure, which could dissipate a large part of the energy input into the structure, as shown in Fig. 1.8. These energy dissipation devices are usually not the main load-bearing of structure, but the structural deformation or vibration will result in the motion or deformation of these devices, and in this way, the devices could dissipate the energy of the structure. Therefore, the dynamic responses and the damages to the main structure will be reduced.

The devices of vibration mitigation mainly include: viscous dampers, viscoelastic dampers, metallic dampers, friction damper, passive tuned dampers, and so on.

1. Metal dampers

    Metal dampers were first researched and tested by Kelly [15], and it takes advantage of the plastic hysteretic deformation of certain metal to dissipate energy. A metal damper mainly includes a mild steel damper, a buckling-restrained brace (BRB) and damper, and so on.

    Many types of mild steel dampers are put forward: typical kinds are X-shape plate dampers and triangle shape dampers. X-shape plate dampers are made by superimposing X-shaped steel plates, and triangle shape dampers are made by

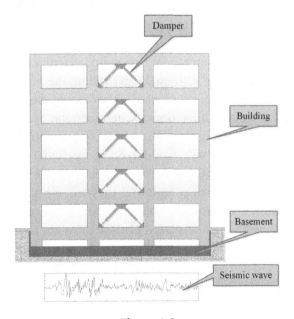

Figure 1.8
Vibration mitigation system.

superposing of multiple triangles steel plates. They dissipate vibration energy of the structure through lateral bending deformation of mild steel. The mild steel damper has a stable hysteretic characteristics, good low cycle fatigue properties, and relative insensitivity to environmental factors. All this indicates that it is a promising damper.

BRB shows the same load-deformation behavior in both compression and tension and higher energy absorption capacity with easy adjustability of both stiffness and strength. Under a minor earthquake, BRB provides additional stiffness for the structure, and under a major earthquake, BRB generates largely plastic hysteretic deformation and dissipates energy.

Lead damper dissipates vibration energy by using extruded or shear hysteretic deformation of lead. As with mild steel dampers, lead dampers have excellent stability and durability properties and insensitivity to environmental factors. But the lead in the damper may cause environmental pollution.

2. Friction dampers

   Friction dampers are generally made up of friction plates, pressure adjustment elements. It dissipates energy through friction forces between the different parts. Friction dampers have rectangular-shape hysteresis curves and a better energy dissipation capacity. Moreover, the damping force can be adjusted by tightening or loosening pressure adjustment elements. However, friction dampers also have some shortcomings, such as the corrosion of the friction surface, which will change the damping force of the damper. The friction damper has no self-reset capability and it must rely on structure stiffness to reset.

3. Viscous dampers

   Viscous damper is mainly composed of the cylinder, guide rods, pistons, damping orifice, and viscous fluid. Under earthquake excitation, the pistons of the viscous damper installed in the structure generates a reciprocating motion, which will force the viscous fluid to flow through the orifices, and then the damping force is generated and the structural dynamic responses are reduced. The viscous fluid damper can be divided into a single rod, a double rod and hydraulic cylinder gap-type forms according to different constructions. Viscous dampers are a kind of velocity-dependent damper. Its distinguishing feature is that only an available additional damping added is into the structure without providing additional stiffness or providing small stiffness, and thus the natural period of the controlled structure is changed slightly. But oil leakage is easy to occur in viscous dampers due to sealing problems. At present, the viscous fluid damper has been used in hundreds of practical engineering cases.

   As for viscous dampers, viscous damping walls are developed [16]. A typical viscous damping wall is made of three plates and high viscosity materials. The internal plate is fixed on the upper floor, two outer plates fixed are on the lower floor. When the structure is subjected to wind or earthquake excitation, the upper and lower floors produce relative motion, and this will lead to the inner plates and the outer plates producing a relative motion and damping force. The viscous damping wall has very good vibration mitigation capacity, but it requires a lot of viscous fluid, which leads to an expensive cost.

4. Viscoelastic dampers

Viscoelastic (VE) dampers consist of VE layers bonded with steel plates by sulfuration. VE dampers dissipate structural vibration energy and reduce structural responses through the shear hysteretic deformation of viscoelastic materials. The hysteresis curves of VE dampers are full ellipses, which indicate that they have excellent energy dissipation capacity. VE dampers not only have advantages of simple construction, easily manufacturing, good durability and low cost, but also have the fine energy dissipation capacity for whatever earthquake or wind excitation. VE dampers can be used both in new constructions for seismic retrofits and in post-earthquake restoration projects. The shortcoming of VE dampers is that the damping properties are affected by temperatures. At present, VE dampers have been used for vibration mitigation of many civil engineering structures.

5. Passive tuned dampers

Passive tuned dampers include the tuned mass damper (TMD) and the tuned liquid damper (TLD). Passive tuned dampers can transfer the energy of the main structure to dampers by tuning frequencies of the controlled main structure, thus mitigating dynamic responses of the main structure.

The TMD system consists of solid mass, springs, and dampers, which is usually installed on the top of the main structure. The mass, stiffness, and damping of this system can be changed to tune the natural period of TMD being same or similar to the natural period of the main structure. When the main structure starts to move, TMD system can generate a force in the opposite direction of the motion of the main structure that acts on the main structure. The TMD system can effectively reduce structural dynamic responses, and the device is of simple construction and is easy to manufacture. The system has little effect on the function of the main structure and is widely used to reduce structural responses induced by earthquake or wind. However, the vibration control effect of the TMD system is mainly affected by the fundamental frequency and the modes of the structure. When the excitation is narrowband or the vibration is controlled by the fundamental mode, the damping effect is excellent; otherwise, the damping effect is worse. That is to say, the TMD system can suppress dynamic responses of the main structure in a narrow range of the excitation frequency.

The TLD system uses water or other liquids as the moving mass has been widely used for vibration mitigation. The basic principle of TLD to dissipate energy of the main structure is the same as TMD. The TLD can use tanks of the structure itself and needs no special device. The TLD gains widespread attention because of convenient application and economy. The TLD also has the advantage of small sloshing damping and the level can be restored after vibration.

Passive tuned dampers are easy to install and maintain. These dampers can be used not only in new structures, but also in existing structures for improving the seismic performance.

1.2.3 Structure Intelligent Control

Though the passive control system has the merits of simple construction, low cost, easy to maintain, not requiring external energy, it usually cannot meet the requirements of precise shock absorption. Intelligent control is usually designed to mitigate vibration responses precisely, and this control technique includes active intelligent control and semi-active intelligent control.

1.2.3.1 Active intelligent control

The active intelligent control system consists of a sensing system, control system, actuator system, and other components. The sensing system measures external disturbances or structural response information and transmits to the control system of the computer. The control system calculates the required control forces based on a given control algorithm and then outputs a control signal to an actuator system. With the help of an external energy source, the actuator system provides the desired control forces on the structure to reduce the responses of the structure [17], as shown in Fig. 1.9. The most common active control devices are active mass damper systems and active tendon control systems.

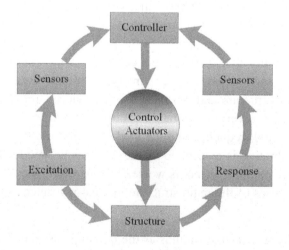

Figure 1.9
Active intelligent control system.

1. Active mass damper system

    An active mass damper (AMD) system is evolved from passive tuned mass damper and provides control forces between inertial mass and structure to adjust the distribution of energy between the main structure and inertial mass. AMD consists of four components: inertial mass, stiffness element, damping element, and the actuator. When AMD begins to work under dynamic loads, the sensors arranged on the structure detect the state of structural responses and translate them to the control system; the control system will calculate the required control forces and send a control signal to the actuators; the control forces will be applied to the structure by actuators through additional mass. An AMD actuator generally adopts the electro-hydraulic servo actuator system or motor servo system and it is the important part of the AMD system for providing active control, which is needed to run to consume large external energy. AMD systems have an excellent mitigation effect in anti-earthquake or wind resistance applications. But the system is sensitive to the stiffness of the structure and the cost is high. This control system can overcome the startup lag problem of passive mass dampers and improve its effectiveness and stability.

2. Active tendon control system

    The active tendon control system can apply control forces on the structure by changing the tendons connected to the structure. The sensors arranged on the structure will detect the structural response state and translate the information to the control system; the control system will calculate the required control forces and send a signal to the actuators; the actuators then adjust the forces of tendons. This system is not sensitive to the stiffness and the damping of the structure, but it is sensitive to the delay. It can provide lateral and tensional control force and has an excellent mitigation effect.

3. Other active control system

    Active support system: This system is formed by adding bracing to a lateral force−resisting member and uses an electro-hydraulic servo mechanism to control bracing deformation. The device is suitable for high-rise and long-span structures. The system is able to make good use of existing structural elements and to minimize cost of the control system.

    Active aerodynamic wind deflector system: This system suppresses structural vibration induced by wind by changing the area of wind shield to adjust suffered wind pressure on the wind shield. As most of the energy comes from wind power, the control system only needs to provide energy for joystick sliding to change the wind shield. This device can be used only to control the vibration induced by wind, and is sensitive to delay.

    Gas pulse generator system: This system generates a control force by a gas shock pulse and the force is a pulse force. The direction and magnitude of pulse control is convenient to adjust. The device is the ideal device for critical state control.

1.2.3.2 Semi-active intelligent control

Semi-active intelligent control systems place the parameters of the structure in an optimal state and control the vibration responses by changing the dynamic characteristics of the main structure, which can be realized by controlling the characteristics of semi-active devices with a small amount of energy. Representative semi-active control devices include a magnetorheological (MR) damper, an electrorheological (ER) damper, a piezoelectric friction damper, a semi-active variable stiffness control device, and a semi-active variable damping control device, etc.

1. Magnetorheological damper

    MR dampers utilize the characteristics of MR fluid, which possesses magnetically controllable yield strength and can reversibly change from free-flow to semi-solid in milliseconds. MR dampers not only have the advantage of a continuously adjustable control force, fast response, low power consumption, but also have an excellent reliability, wide dynamic range, high frequency response. Therefore, various forms of MR dampers have been proposed [18,19]. MR dampers mainly include forms of squeeze flow type, shear type, valve type, and shear valve type. MR dampers are ideal devices for reducing structural responses induced by wind or earthquake, and there are good prospects for development of MR dampers.

2. Electrorheological damper

    ER damper utilizes another kind of controllable fluid: ER fluid. The damping force provided by the device can be adjusted by changing the properties of the fluid through applying a different electric field strength. Similar to MR damper, ER dampers also have the advantage of a continuously adjustable control force, fast response, low power consumption, excellent reliability, wide dynamic range, and a high frequency response. ER dampers also mainly include forms of squeeze flow type, shear type, valve type, and the shear valve type. However, the shear yield strength of the ER fluid is too low and the required electric field strength is too high.

3. Piezoelectric friction damper

    The piezoelectric friction damper is developed based on friction dampers. This damper can regulate the friction force by adjusting the positive pressure on the contact surface through piezoelectric actuators. Piezoelectric actuators can adjust pressures in real time in accordance with vibration states by using the converse piezoelectric effect of piezoelectric material [20]. Piezoelectric friction dampers have the advantages of self-adaptive vibration control like other semi-active control devices. However, this damper has shortcomings of friction dampers, such as instability of the friction parameters.

4. Semi-active variable stiffness damper

    Semi-active variable stiffness damper consists of a stiffness element and an intelligent control component. When a semi-active variable stiffness damper changes, the additional real-time stiffness, in accordance with vibration states, reduces structural responses self-adaptively. The natural frequency of the system is away from the predominant frequency of excitations to avoid the structural resonance when the structural stiffness is changed in real time. Semi-active variable stiffness dampers are adaptive to the structures with small stiffness. For the large-scaled structures, the changes of stiffness due to the dampers are usually small compared with the main structures. Therefore, the adjustment range of natural frequency of the structure is narrow.

5. Semi-active variable damping damper

   Semi-active variable damping damper mainly consists of a damping element and an intelligent control component. This kind of damper can provide an appropriate damping force to reduce the responses of the main structure by adjusting the additional damping in real time. Semi-active variable damping controls are unconditionally stable and have good robustness, because damping is always beneficial to vibration mitigation for structures.

6. Magnetorheological elastomer device

   Apart from MR fluid, MR elastomer is mainly the mixture of polymers (such as rubber, etc.) and soft magnetic particles (such as carbonyl iron powder). Because the particles are mixed into the base rubber, there is no particle sedimentation problem. Therefore, compared with the ordinary MR fluid, MR elastomer not only has merits of controllable, reversible, fast-response characteristics, but also has advantages of fine stability, simple design, and low cost. MR elastomer device is made of MR elastomer and magnetic field adjustment part. Currently, the new developed intelligent control device has attracted increased attention.

7. Shape memory alloy dampers

   Shape memory alloy (SMA) is a kind of special metallic materials which can change its shape and has a reversible change when the external environment conditions change. SMA damper dissipates vibration energy by using the shape memory effect and super-elastic behavior of SMA. SMA damper has the merits of high damping, high resilience, and driving characteristics; it is a widely used smart material. It also has good fatigue and corrosion resistance and durability. But the properties of SMA may be sensitive to temperature. Although it has some disadvantages, it is still an ideal material used in vibration control.

8. Other semi-active control devices

   Besides the above-mentioned semi-active control devices, there are some other semi-active control devices, such as semi-active TMD and semi-active TLD. These semi-active control devices all have self-adjustable characteristics and can achieve adjustable control forces with much less energy than active control.

1.2.3.3 Intelligent control algorithm

The intelligent control algorithm can be considered as a mathematical object to be formulated within the framework of a control theory. The algorithm is designed to a controller, which determines the parameters of semi-active control devices or forces of active control devices in accordance with the measured vibration state or excitation forces. A control algorithm is very important in intelligent control and is one of the main factors deciding control precision. The two types of control strategy are realized using feedback and feed forward controllers, respectively. At present, the main control strategies proposed for intelligent control are: classical linear optimal control algorithm, pole assignment method, instant optimal control algorithm, independent mode space control, H_∞ feedback control, slipping mode control, optimal polynomial control, fuzzy control, neural network fuzzy control, particle swarm control algorithm, and genetic algorithm. The introduction to these control algorithms will be described in Chapter 2, Intelligent Control Strategies.

CHAPTER 2

Intelligent Control Strategies

For the intelligent vibration control system, the control algorithm is the core aspect to realize vibration mitigation effectively. In this chapter, equations of motion of an intelligent vibration control system are introduced firstly, then some typical control algorithms are introduced, including classical linear optimal control algorithm, pole assignment method, instantaneous optimal control algorithm, independent mode space control (IMSC), H_∞ feedback control, sliding mode control (SMC), optimal polynomial control (OPC), fuzzy control, neural network, fuzzy control, particle swarm control algorithm, and genetic algorithm (GA).

2.1 Equations of Motion of Intelligent Control System

In Chapter 1, Introduction, the equation of motion of civil structures with energy dissipation devices under dynamic loads has been given as Eq. (1.4). When the energy dissipation devices are the intelligent dampers, the equation of motion of the structural system with a finite number n of degrees of freedom under dynamic loads can be expressed as:

$$[M]\{\ddot{x}(t)\} + [C]\{\dot{x}(t)\} + [K]\{x(t)\} = \{P(t)\} + [B]\{f_d(t)\} \quad (2.1)$$

where $\{\ddot{x}(t)\}$, $\{\dot{x}(t)\}$, $\{x(t)\}$ are, respectively, the vectors of acceleration, velocity, and displacement responses. $\{P(t)\}$ is the external excitation load vector, such as, the earthquake acceleration or the wind load. $[B]$ is the position matrix. $\{f_d(t)\}$ is the control force vector provided by intelligent dampers, and it is calculated according to some different control algorithms. $[M]$ is the structural lumped mass matrix.

$$[M] = \begin{bmatrix} m_1 & 0 & \cdots & 0 \\ 0 & m_2 & \cdots & 0 \\ \cdots & \cdots & \cdots & \cdots \\ 0 & 0 & \cdots & m_n \end{bmatrix}$$

$[K]$ is the structural stiffness matrix

$$[K] = \begin{bmatrix} k_{11} & k_{12} & \cdots & k_{1n} \\ k_{21} & k_{22} & \cdots & k_{2n} \\ \cdots & \cdots & \cdots & \cdots \\ k_{n1} & k_{n2} & \cdots & k_{nn} \end{bmatrix}$$

$[C]$ is the structural damping matrix, and it is usually calculated in accordance with the most popular hypotheses damping, the *Rayleigh damping*:

$$[C] = \kappa_1[M] + \kappa_2[K] \qquad (2.2)$$

The coefficients κ_1 and κ_2 are selected to fit the structure under consideration. That is,

$$\kappa_1 = \frac{2\omega_1\omega_2(\zeta_1\omega_2 - \zeta_2\omega_1)}{\omega_2^2 - \omega_1^2} \quad \kappa_2 = \frac{2(\zeta_2\omega_2 - \zeta_1\omega_1)}{\omega_2^2 - \omega_1^2} \qquad (2.3)$$

where ω_1, ω_2, ζ_1, ζ_2 are the first-, the second-order frequencies, and the corresponding damping ratios.

If the system is a building structure and the external excitation load $\{P(t)\}$ is the earthquake acceleration, Eq. (2.1) can be rewritten as:

$$[M]\{\ddot{x}(t)\} + [C]\{\dot{x}(t)\} + [K]\{x(t)\} = -[M]\{\Gamma\}\ddot{x}_g(t) + [B]\{f_d(t)\} \qquad (2.4)$$

where $\{\Gamma\} = [1, 1, \ldots 1]^T$ is a column vector of ones; $\ddot{x}_g(t)$ is the earthquake acceleration excitation.

In control applications, Eq. (2.1) is always rewritten as the state-space equations:

$$\begin{cases} \{\dot{Z}(t)\} = [A]\{Z(t)\} + [D_0]\{P(t)\} + [B_0]\{f_d(t)\} \\ \{Y\} = [E_0]\{Z(t)\} \end{cases} \qquad (2.5)$$

where $\{Z(t)\} = \{\{x(t)\}, \{\dot{x}(t)\}\}^T$ is the state vector; $\{Y\}$ is the output vector; $[A] = \begin{bmatrix} [0] & [I] \\ -[M]^{-1}[K] & -[M]^{-1}[C] \end{bmatrix}$ is the system matrix; $[D_0] = \begin{bmatrix} [0] \\ [M]^{-1} \end{bmatrix}$; $[B_0] = \begin{bmatrix} [0] \\ [M]^{-1}[B] \end{bmatrix}$ is the input matrix; $[E_0] = \begin{bmatrix} [I] & [0] \end{bmatrix}$ is the output matrix, where $[0]$ is the null matrix, $[I]$ is the identity matrix.

2.2 Classical Linear Optimal Control Algorithm

The classical linear optimal control algorithm is one of the branches of modern control theories. It is applied in many fields such as national defense, physics, chemistry, economical management, and social science. At present, the most popular linear optimal control is the linear quadratic optimal control, including linear quadratic regulator (LQR) and linear quadratic gauss (LQG) control algorithms. In the linear quadratic optimal control, the integration of quadratic functions, which contains the system state and the control input vector, is proposed as the performance index function (PIF). Based on the PIF, the solution of the optimal control problem can be obtained.

2.2.1 LQR Optimal Control

The LQR optimal control method has drawn more attention due to its effectiveness and convenience. For a certain problem, LQR optimal control can be realized by adjusting weights matrices.

2.2.1.1 Basic equation of LQR optimal control

Based on Eq. (2.5), the state-space equation of a controllable linear time-invariant system without external excitation can be written as:

$$\{\dot{Z}(t)\} = [A]\{Z(t)\} + [B_0]\{f_d(t)\} \tag{2.6a}$$

$$\{Y(t)\} = [E_0]\{Z(t)\} \tag{2.6b}$$

when the initial state of the system is defined as $\{Z_0\}$, $\{Z_0\} = \{Z(t_0)\}$.

Choose the quadratic PIF as follows:

$$J = \int_{t_0}^{\infty} [\{Z(t)\}^T[Q]\{Z(t)\} + \{f_d(t)\}^T[R]\{f_d(t)\}] \, dt \tag{2.7}$$

where $[Q]$ and $[R]$ are weighting matrices, $[Q]$ is a positive semi-definite matrix, and $[R]$ is a positive-definite matrix.

In order to minimize J, the Lagrange multiplier factor method is introduced with the factor $\{\tilde{\lambda}(t)\}$. The Lagrange function has the form

$$\{L\} = \int_{t_0}^{\infty} \begin{bmatrix} \{Z(t)\}^T[Q]\{Z(t)\} + \{f_d(t)\}^T[R]\{f_d(t)\} \\ + \{\tilde{\lambda}\}^T([A]\{Z(t)\} + [B_0]\{f_d(t)\} - \{\dot{Z}(t)\}) \end{bmatrix} dt \tag{2.8}$$

2.2.1.2 Solution of optimal control

Based on the classical variation method [21], Eq. (2.8) can be rewritten as:

$$\{L\} = \int_{t_0}^{\infty} \left[H(\{Z\}, \{f_d\}, \{\tilde{\lambda}\}, t) + \left\{\dot{\tilde{\lambda}}\right\}^T \{Z\} \right] dt - \{\tilde{\lambda}\}^T\{Z\}\Big|_{t_0}^{\infty} \tag{2.9}$$

where

$$\{H(\{Z\},\{f_d\},\{\tilde{\lambda}\},t)\} = \{Z(t)\}^T[Q]\{Z(t)\} + \{f_d(t)\}^T[R]\{f_d(t)\} \\ + \{\tilde{\lambda}\}^T([A]\{Z(t)\} + [B_0]\{f_d(t)\}) \tag{2.10}$$

Considering the first-order variation, the following equation can be obtained

$$\delta\{L\} = \delta\{L_Z\} + \delta\{L_{f_d}\} + \delta\{L_{\tilde{\lambda}}\} \\ = \int_{t_0}^{\infty} \left[\delta\{Z\}^T \left(\frac{\partial\{H\}}{\partial\{Z\}} + \{\dot{\tilde{\lambda}}\} \right) + \delta\{f_d\}^T \frac{\partial\{H\}}{\partial\{f_d\}} + \delta\{\tilde{\lambda}\}^T \left(\frac{\partial\{H\}}{\partial\{\tilde{\lambda}\}} - \{\dot{Z}\} \right) \right] dt - \delta\{Z\}^T\{\tilde{\lambda}\} \Big|_{t_0}^{\infty} \tag{2.11}$$

in which,

$$\delta\{L_Z\} = \delta\{Z\}^T \frac{\partial\{L\}}{\partial\{Z\}} = \int_{t_0}^{\infty} \delta\{Z\}^T \left(\frac{\partial\{H\}}{\partial\{Z\}} + \{\dot{\tilde{\lambda}}\} \right) dt - \delta\{Z\}^T\{\tilde{\lambda}\} \Big|_{t_0}^{\infty}$$

$$\delta\{L_{f_d}\} = \delta\{f_d\}^T \frac{\partial\{L\}}{\partial\{f_d\}} = \int_{t_0}^{\infty} \delta\{f_d\}^T \frac{\partial\{H\}}{\partial\{f_d\}} dt$$

$$\delta\{L_{\tilde{\lambda}}\} = \delta\{\tilde{\lambda}\}^T \frac{\partial\{L\}}{\partial\{\tilde{\lambda}\}} = \int_{t_0}^{\infty} \delta\{\tilde{\lambda}\}^T \left(\frac{\partial\{H\}}{\partial\{\tilde{\lambda}\}} - \{\dot{Z}\} \right) dt$$

In order to satisfy the requirements of the extreme value condition, $\delta\{L\} = 0$, with the arbitrary value of $\delta\{Z\}$, $\delta\{f_d\}$, and $\delta\{\tilde{\lambda}\}$, the following equations can be obtained

$$\frac{\partial\{H\}}{\partial\{Z\}} + \{\dot{\tilde{\lambda}}\} = 0, \frac{\partial\{H\}}{\partial\{f_d\}} = 0, \frac{\partial\{H\}}{\partial\{\tilde{\lambda}\}} - \{\dot{Z}\} = 0, \delta\{Z\}^T\{\tilde{\lambda}\} \Big|_{t_0}^{\infty} = 0 \tag{2.12}$$

Based on Eqs. (2.9)–(2.12), the following equations can be obtained

$$\{\dot{\tilde{\lambda}}\} = \frac{-\partial\{H\}}{\partial\{Z\}} = -2[Q]\{Z\} - [A]^T\{\tilde{\lambda}\} \tag{2.13}$$

$$\frac{\partial\{H\}}{\partial\{f_d\}} = 2[R]\{f_d\} + [B_0]^T\{\tilde{\lambda}\} = 0 \tag{2.14}$$

Eq. (2.14) is rewritten as:

$$\{f_d(t)\} = -\frac{1}{2}[R]^{-1}[B_0]^T\{\tilde{\lambda}(t)\} \tag{2.15}$$

Then it can be concluded that $\{f_d(t)\}$ is a linear function of $\{\tilde{\lambda}(t)\}$. The linear transform relation can be assumed as

$$\{\tilde{\lambda}(t)\} = \{\tilde{P}(t)\}\{Z(t)\} \tag{2.16}$$

Substituting Eq. (2.16) into Eq. (2.15), let $[G] = \frac{1}{2}[R]^{-1}[B_0]^T\{\tilde{P}(t)\}$, it can be obtained as

$$\{f_d(t)\} = -[G]\{Z(t)\} \tag{2.17}$$

Substituting Eq. (2.17) into Eq. (2.13), it can be obtained as

$$\{\dot{\tilde{P}}(t)\}\{Z(t)\} + \{\tilde{P}(t)\}\{\dot{Z}(t)\} = -(2[Q] + [A]^T\{\tilde{P}(t)\})\{Z(t)\} \tag{2.18}$$

Simplifying Eq. (2.18) with Eqs. (2.6) and (2.15), it can be obtained as

$$\{\dot{\tilde{P}}(t)\} + \{\tilde{P}(t)\}[A] + [A]^T\{\tilde{P}(t)\} - \frac{1}{2}\{\tilde{P}(t)\}[B_0][R]^{-1}[B_0]^T\{\tilde{P}(t)\} + 2[Q] = 0 \tag{2.19}$$

Eq. (2.19) can be rewritten as:

$$\{\dot{\tilde{P}}(t)\} = -\{\tilde{P}(t)\}[A] - [A]^T\{\tilde{P}(t)\} + \frac{1}{2}\{\tilde{P}(t)\}[B_0][R]^{-1}[B_0]^T\{\tilde{P}(t)\} - 2[Q] \tag{2.20}$$

Eq. (2.20) is called the Riccati equation, which is a nonlinear matrix differential equation, $\{\tilde{P}(t)\}$ the Riccati matrix function. The solution of Eq. (2.20) can be calculated by the method shown in references [22,23].

Substituting Eq. (2.17) into Eq. (2.6), it can be obtained as

$$\{\dot{Z}(t)\} = \left([A] - \frac{1}{2}[B_0][R]^{-1}[B_0]^T[\tilde{P}(t)]\right)\{Z(t)\} \tag{2.21}$$

LQR studies the linear system in a state-space form given by modern control theory. The PIF is based on the system state and the control force. The PIF J reckons on the matrices $[Q]$ and $[R]$: the assignment of large value to the element of $[Q]$ indicates the reduction of the control response and the better control effect, simultaneously; on the contrary, the control effect will be not obvious when the elements of $[R]$ are large in comparison with those of $[Q]$.

2.2.2 LQG Optimal Control

LQG is another kind linear, quadratic optimal control algorithm. The state feedback control $\{f_d(t)\}$ is obtained using LQR. In this method, Kalman filter [24–26] is applied to estimate the system state approximately. For a nonlinear discrete-time controlled process of the form [24–26]

$$\begin{cases} \{Z(t)_{k+1}\} = f(\{Z(t)_k\}, \{f_d(t)_k\}, \{\nu(t)\}, k) \\ \{Y_k\} = h(\{Z(t)_k\}, \{f_d(t)_k\}, k) + \{\omega(t)\} \end{cases} \quad (2.22)$$

where $\{Z(t)_k\}$ is the state of the system at time step k, $\{f_d(t)_k\}$ is the input vector, $\{\nu(t)\}$ is the state noise process vector due to disturbances and modeling errors, $\{Y_k\}$ is the observation vector, and $\{\omega(t)\}$ is the measurement noise. It is assumed that both $\{\nu(t)\}$ and $\{\omega(t)\}$ are Gaussian white noise with zero mean and $E[\{\nu(t)\}] = 0$; $E[\{\omega(t)\}] = 0$; $E[\{\nu(t)\}\{\nu^T(\tau)\}] = Q_e \delta(t-\tau), Q_e = Q_e^T \geq 0$.

$E[\{\omega(t)\}\{\omega^T(\tau)\}] = R_e \delta(t-\tau), R_e = R_e^T \geq 0$, where $E[\cdot]$ is a mean value. Q_e and R_e are the covariance of $\{\nu(t)\}$ and $\{\omega(t)\}$.

When the Kalman filter is applied to the system shown as Eq. (2.5) without external excitation, the state-space equation can be rewritten as follows:

$$\{\dot{Z}(t)\} = [A]\{Z(t)\} + [B_0]\{f_d(t)\} + \{\nu(t)\} \quad \{Z(t_0)\} = \{Z_0\} \quad (2.23a)$$

$$\{Y(t)\} = [E_0]\{Z(t)\} + \{\omega(t)\} \quad (2.23b)$$

The full state feedback optimal control $\{f_d(t)\}$ should be determined by LQR optimal control algorithm, as shown in Eq. (2.17), $\{f_d(t)\} = -[G]\{Z(t)\}$. Due to the basic theory of Kalman filter [24–26], for a linear relation in Eq. (2.23b), the Kalman filter gain matrix $[K_e]$ has the form

$$[K_e] = [P_e][E_0]^T [R_e]^{-1} \quad (2.24)$$

$[P_e] = \lim_{t \to \infty} E[\{\{Z(t)\} - \{\hat{Z}(t)\}\}\{\{Z(t)\} - \{\hat{Z}(t)\}\}^T]$. $[P_e]$ is the covariance matrix of the steady-state error of the filter. $[P_e]$ is the solution of the Riccati equation in Eq. (2.25) [27]

$$[P_e][A]^T + [A][P_e] - \frac{1}{2}[P_e][E_0]^T [R_e]^{-1} [E_0][P_e] + 2[Q_e] = 0 \quad (2.25)$$

Then, based on the observed output, the full state of the structure is estimated by the Kalman filter. The objective function is chosen as

$$\{J_e\} = E[\{\{Z(t)\} - \{\hat{Z}(t)\}\}^T \{\{Z(t)\} - \{\hat{Z}(t)\}\}] \quad (2.26)$$

$$\{\hat{Y}\} = [E_0]\{\hat{Z}\} \quad (2.27)$$

In the above equations, $\{\hat{Z}(t)\}$ is the estimation of the state $\{Z(t)\}$, then the Kalman filter has the form

$$\{\dot{\hat{Z}}\} = [A]\{\hat{Z}\} + [B_0]\{f_d(t)\} + [K_e](\{Y\} - \{\hat{Y}\}) \quad \{\hat{Z}(t_0)\} = \{\hat{Z}_0\} \tag{2.28}$$

$$\{\hat{Y}\} = [E_0]\{\hat{Z}\} \tag{2.29}$$

At last, the optimal control with state estimate feedback can be obtained

$$\{f_d(t)\} = -[G]\{\hat{Z}(t)\} \tag{2.30}$$

Substituting Eq. (2.30) into Eq. (2.28), the system state can be written as:

$$\{\dot{\hat{Z}}\} = ([A] - [B_0][G] - [K_e][E_0])\{\hat{Z}\} + [K_e]\{Y\} \quad \{\hat{Z}(t_0)\} = \{\hat{Z}_0\} \tag{2.31}$$

$$\{\hat{Y}\} = [E_0]\{\hat{Z}\} \tag{2.32}$$

LQG is not a full state control which needs to feedback the whole state information of the system. In many specific circumstances with a large data volume, especially for mega structures, it is not executable and economical to measure the full state variables of the system. So the local observation and state estimation are employed to realize the feedback control. LQG optimal control is an approximately estimation about the structure, and the asymptotic value is accurate enough for the engineering requirement.

The minimization procedure in the optimal linear control does not take the external excitation into account. Therefore this control algorithm is not truly optimal. In fact, many control algorithms are not truly optimal in this sense. The optimal linear control can be given using an open–close loop structure incorporating the measured excitation feedback. Many studies have been carried out on linear optimal control [28–33].

2.3 Pole Assignment Method

The pole assignment method is usually employed for controlling the system without external excitation. The pole of a system is the eigenvalue of the system matrix $[A]$. The eigenvalue could be real or complex. The complex eigenvalue refers to a point in the complex plane. To a great extent, the dynamic responses of the system are determined by the complex plane position of the pole. Using state/output feedback, the damping matrix and the stiffness matrix will be changed, as well as the pole position. This process to achieve control of the system is defined as the pole assignment method.

In the absence of systematic disturbance (external excitation), the pole assignment method for the linear time-invariant system as below is discussed

$$\{\dot{Z}(t)\} = [A]\{Z(t)\} + [B_0]\{f_d(t)\} \tag{2.33}$$

$$\{Y(t)\} = [E_0]\{Z(t)\} \tag{2.34}$$

2.3.1 Pole Assignment Method with State Feedback

When the pole assignment method with state feedback is utilized in the system, assume that the input vector $\{f_d(t)\}$ in Eq. (2.33) has this form

$$\{f_d(t)\} = -[G]\{Z(t)\} \tag{2.35}$$

where $[G]$ is the state feedback gain matrix.

Considering Eqs. (2.33) and (2.35), it can be obtained

$$\{\dot{Z}(t)\} = ([A] - [B_0][G])\{Z(t)\} \tag{2.36}$$

For a closed-loop system in Eq. (2.36), it is obvious that

$$\begin{aligned}\Delta(\lambda) &= |(\lambda[I_0]) - ([A] - [B_0][G])| \\ &= |\lambda[I_0] - [A]| \cdot |[I_0] + (\lambda[I_0] - [A])^{-1}[B_0][G]|\end{aligned} \tag{2.37}$$

where $[I_0]$ is an unit matrix of arbitrary order. λ is the eigenvalue of the closed-loop system.

Based on Eq. (2.37), there is

$$\begin{aligned}\Delta(\lambda) &= |\lambda[I_0] - [A]| \cdot |[I_0] + (\lambda[I_0] - [A])^{-1}[B_0][G]| \\ &= |\lambda[I_0] - [A]| \cdot |[I_0] + [G](\lambda[I_0] - [A])^{-1}[B_0]| = 0\end{aligned} \tag{2.38}$$

As known, $|\lambda[I_0] - [A]| \neq 0$, there is

$$\left|[I_0] + [G](\lambda[I_0] - [A])^{-1}[B_0]\right| = 0 \tag{2.39}$$

From Eq. (2.39), the appropriate option of the gain matrix is obtained [27].

$$[G] = -[e][\Gamma]^{-1} \tag{2.40}$$

where $[\Gamma]$ is made up of the eigenvalues of the closed-loop system, $[e]$ is a matrix derived from the unit matrix.

2.3.2 Pole Assignment Method With Output Feedback

The pole assignment method with output feedback could be applied when the system is observable and controllable.

Assuming that the input vector $\{f_d(t)\}$ in Eq. (2.33) has this form

$$\{f_d(t)\} = -[G']\{Y(t)\} \tag{2.41}$$

Substituting Eq. (2.41) into Eq. (2.33), it can be obtained

$$\{f_d(t)\} = -[G'][E_0]\{Z(t)\} \tag{2.42}$$

Based on Eq. (2.33) and Eq. (2.42), it can be obtained

$$\{\dot{Z}(t)\} = ([A] - [B_0][G'][E_0])\{Z(t)\} \tag{2.43}$$

For a closed-loop system in Eq. (2.43),

$$\Delta(\lambda) = \left|(\lambda[I_0]) - ([A] - [B_0][G'][C_0])\right| = \left|\lambda[I_0] - [A]\right| \cdot \left|[I_0] + [G'][\varphi'(\lambda)]\right| = 0 \tag{2.44}$$

where

$$[\varphi'(\lambda)] = [C_0](\lambda[I_0] - [A])^{-1}[B_0] \tag{2.45}$$

For a closed-loop system, $|\lambda[I_0] - [A]| \neq 0$, and it can be obtained

$$\left\|[I_0] + [G'][\varphi'(\lambda)]\right\| = 0 \tag{2.46}$$

As the same way in Eq. (2.40), the expression of output feedback gain matrix is obtained [27]

$$[G'] = -[\varepsilon_1][\Gamma_1]^{-1} \tag{2.47}$$

where $[\Gamma_1]$ and $[\varepsilon_1]$ are new matrices derived from the eigenvalue of the closed-loop system and the unit matrix.

As seen in Fig. 2.1, the process of system control with pole assignment method can be described as: step I, assume the system state, and build the relationship between the input force and the system state; step II, obtain the state/output feedback gain matrix with pole assignment method; and step III, after observation and calculation, the expression of gain matrix and system state are obtained.

There have been a lot of researches on pole assignment method in the field of system control [34,35]. However, only few modes of vibration are considered when pole

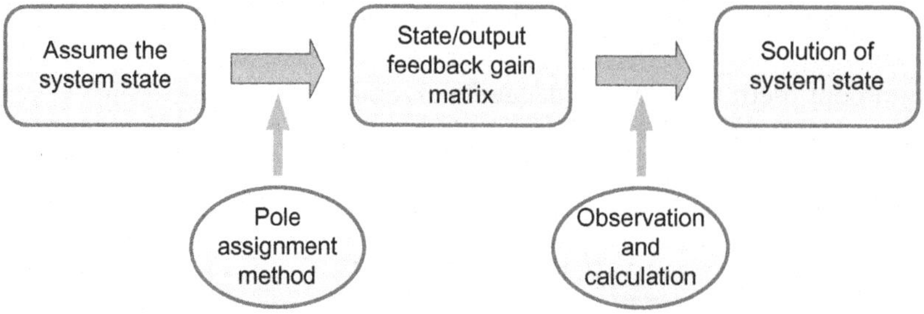

Figure 2.1
The process of system control design with pole assignment method.

assignment method is applied in the structural vibration control [36,37]. This also shows pole assignment method is effective in a system with the only few modes of vibration as predominant. It is noted that some optimality criteria are required to be incorporated into the pole assignment process because the gain matrix [G] based on prescribed eigenvalues is in general not unique. Hence some optimality criteria are required to be incorporated into the pole assignment process.

2.4 Instantaneous Optimal Control Algorithm

The instantaneous optimal control method was first put forward by Yang [38–40] based on classical optimal control algorithms. The closed-loop optimal control is not truly optimal because the external excitation is ignored. Recognizing that, at any particular time t, the influence of external excitation may be available up to the time instant t, which could be utilized in the developed control algorithm. These improved control laws are referred to as the instantaneous optimal control algorithm [22,23].

In accordance with Eq. (2.5), the state-space equation of a controllable linear time-invariant system with external load can be expressed as

$$\begin{cases} \{\dot{Z}(t)\} = [A]\{Z(t)\} + [B_0]\{f_d(t)\} + [D_0]\{P(t)\} \quad \{Z(t_0)\} = \{Z_0\} \\ \{Y(t)\} = [E_0]\{Z(t)\} \end{cases} \quad (2.48)$$

where $\{Z(t_0)\} = \{Z_0\}$ is the initial system state, t is an instant time.

The instantaneous optimal control will be established with time-dependent PIF $J(t)$

$$J(t) = \{Z(t)\}^T[Q]\{Z(t)\} + \{f_d(t)\}^T[R]\{f_d(t)\} \quad (2.49)$$

Decouple the first equation of Eq. (2.48) with the transformation below

$$\{Z(t)\} = [T]\{Y_1\} \tag{2.50}$$

where $[T]$ is a model matrix consists of eigenvectors of $[A]$.

Substituting Eq. (2.50) into Eq. (2.48), there will be

$$\begin{aligned}\{\dot{Y}_1(t)\} &= [\lambda]\{Y_1(t)\} + \{\tilde{F}(t)\} \\ \{Y_1(0)\} &= \{0\} \quad \{Y(t)\} = [E_0]\{Z(t)\}\end{aligned} \tag{2.51}$$

where $[\lambda]$ is a diagonal matrix consisting of complex eigenvalues $\lambda_j, (j = 1, 2, \ldots, n)$ of matrix $[A]$, and

$$\{\tilde{F}(t)\} = [T]^{-1}\{[B_0]\{f_d(t)\} + [D_0]\{P(t)\}\} \tag{2.52}$$

The solution of Eq. (2.51) has the form

$$\{Y_1(t)\} = \int_0^t \exp[\lambda(t-\tau)]\tilde{F}(\tau)d\tau \tag{2.53}$$

where $\exp[\lambda(t-\tau)]$ is a diagonal matrix with the jth diagonal element being $\exp[\lambda_j(t-\tau)]$. Eq. (2.53) can be rewritten as:

$$\begin{aligned}\{Y_1(t)\} &= \int_0^{t-\Delta t} \exp[\lambda(t-\tau)]\tilde{F}(\tau)d\tau + \int_{t-\Delta t}^t \exp[\lambda(t-\tau)]\tilde{F}(\tau)d\tau \\ &= e^{[\lambda(\Delta t)]}Y_1(t-\Delta t) + \int_{t-\Delta t}^t \exp[\lambda(t-\tau)]\tilde{F}(\tau)d\tau \\ &= e^{[\lambda(\Delta t)]}Y_1(t-\Delta t) + \frac{\Delta t}{2}e^{[\lambda(\Delta t)]}[\tilde{F}(t-\Delta t) + \tilde{F}(t)]\end{aligned} \tag{2.54}$$

Substituting Eq. (2.54) into Eq. (2.50), there will be

$$\begin{aligned}\{Z(t)\} &= [T]e^{[\lambda(\Delta t)]}Y_1(t-\Delta t) + \frac{\Delta t}{2}[T]e^{[\lambda(\Delta t)]}[\tilde{F}(t-\Delta t) + \tilde{F}(t)] \\ &= [T]e^{[\lambda(\Delta t)]}[T]^{-1}Z(t-\Delta t) + \frac{\Delta t}{2}[T]e^{[\lambda(\Delta t)]}[\tilde{F}(t-\Delta t) + \tilde{F}(t)]\end{aligned} \tag{2.55}$$

Substituting Eq. (2.52) into Eq. (2.55), it can be obtained

$$\{Z(t)\} = [T]\{\tilde{D}(t-\Delta t)\} + \frac{\Delta t}{2}\{[B_0]\{f_d(t)\} + [D_0]\{P(t)\}\} \tag{2.56}$$

In the above equation

$$\{\tilde{D}(t-\Delta t)\} = e^{[\lambda(\Delta t)]}[T]^{-1}\left\{\{Z(t-\Delta t)\} + \frac{\Delta t}{2}\left\{\begin{array}{l}[B_0]\{f_d(t-\Delta t)\}\\+[D_0]\{P(t-\Delta t)\}\end{array}\right\}\right\} \quad (2.57)$$

In order to obtain the optimal extreme value with the Lagrange multiplier factor method mentioned in Section 2.2.1, a new function $H(t)$ is defined

$$\begin{aligned}\{H(t)\} &= \{Z(t)\}^T[Q]\{Z(t)\} + \{f_d(t)\}^T[R]\{f_d(t)\}\\&+ \{\tilde{\lambda}\}\left\{\begin{array}{l}\{Z(t)\} - [T]\{\tilde{D}(t-\Delta t)\}\\-\frac{\Delta t}{2}\{[B_0]\{f_d(t)\} + [D_0]\{P(t)\}\}\end{array}\right\}\end{aligned} \quad (2.58)$$

The relationship between the control vector $\{f_d(t)\}$ and the system state space $\{Z(t)\}$ is assumed

$$\{\tilde{\lambda}(t)\} = \{\tilde{P}(t)\}\{Z(t)\} \quad (2.59)$$

Then based on the variation method [21], as the same way in Section 2.2.1, the control vector $\{f_d(t)\}$ for the instantaneous closed-loop system will be obtained

$$\{f_d(t)\} = -\frac{\Delta t}{2}[R]^{-1}[B_0]^T[Q]\{Z(t)\} \quad (2.60)$$

The response state vector $\{Z(t)\}$ under the optimal closed-loop control is determined from Eq. (2.56) with the aid of Eq. (2.60) as follows:

$$\{Z(t)\} = \left[[I_0] + \frac{(\Delta t)^2}{2}[B_0][R]^{-1}[B_0]^T[Q]\right]^{-1}\left\{\begin{array}{l}[T]\{\tilde{D}(t-\Delta t)\}\\+\frac{\Delta t}{2}[D_0]\{P(t)\}\end{array}\right\} \quad (2.61)$$

The time interval and weight matrix have strong effects on dynamic responses of the controlled structures. Depending on the selection of weight matrices $[Q]$ and $[R]$, various control effectiveness could be achieved. Furthermore, classical optimal control algorithms require the solution of the Riccati equation in Eq. (2.20), which can be cumbersome for a structure with many degrees of freedom, while instantaneous optimal control can be employed without the solution of Riccati equation.

2.5 Independent Mode Space Control

IMSC is a common and simple structural vibration control method, which mainly combines the modal decomposition and some control algorithms [41]. Generally, the structural dynamic responses are dominated by a small number of particular modes of structures, and the structural dynamic analysis results can be achieved with sufficient accuracy by selecting some dominated modes. The modal control force can be calculated in accordance with common control algorithms, which has the merit of simple calculation due to the limited controlled modes [42].

2.5.1 Modal Control Based on State Space

The state-space equation of a controlled linear time-invariant system can be expressed as:

$$\{\dot{Z}(t)\} = [A]\{Z(t)\} + [B_0]\{f_d(t)\} \quad (2.62)$$

where the system matrix $[A]$ has n pairs of complex conjugate eigenvalues, and the jth pair of eigenvalues are expressed as $-\beta_j \pm iw_j$. Accordingly, the n pairs of eigenvectors are also complex conjugates, and the jth pair of eigenvectors are $a_j \pm ib_j$.

The eigenvectors and eigenvalues of the controlled system can be written in the matrix form: $[\Psi] = [a_1 b_1 \vdots a_2 b_2 \vdots \cdots \vdots a_n b_n]$, $[\Lambda_j] = \begin{bmatrix} \beta_j & w_j \\ -w_j & \beta_j \end{bmatrix}$. Considering that these eigenvectors are orthogonal with respect to the matrix $[A]$, the relation between eigenvalues and eigenvectors is expressed as:

$$[\Lambda] = [\Psi]^{-1}[A][\Psi] \quad (2.63)$$

Then the status of the controlled system can be expressed with conjugate eigenvectors $[\Psi]$:

$$\{Z(t)\} = [\Psi]\{\xi(t)\} \quad (2.64)$$

where $\{\xi(t)\}$ is the generalized coordinate vector of the controlled system. Based on the above orthogonal condition, Eq. (2.62) can be rewritten as:

$$\{\dot{\xi}(t)\} = [\Lambda]\{\xi(t)\} + \{f_m(t)\} \quad (2.65)$$

$$\{f_m(t)\} = \{\{f_{m1}\}^T \{f_{m2}\}^T \cdots \{f_{mn}\}^T\}^T = [\Psi]^{-1}[B_0]\{f_d(t)\} \quad (2.66)$$

where $\{f_m(t)\}$ is the generalized control force vector. Thus this operation converts the coupled equation of the system to n pairs of independent equations, and the jth pair of equations can describe the control of the jth pair of modes, the detailed equation is:

$$\{\dot{\xi}_j(t)\} = [\Lambda_j]\{\xi_j(t)\} + \{f_{mj}(t)\} \quad (2.67)$$

Thus a control problem of $2n$-dimensional coupled system is transformed into control problems of the n decoupled systems. The modal control force $\{f_{mj}(t)\}$ can be calculated by a particular control algorithm:

$$\{f_{mj}(t)\} = -[G_j]\{\xi_j(t)\} \tag{2.68}$$

Assuming there are n_c modes of system to be controlled, Eq. (2.68) can be rewritten as:

$$\{f_{mc}(t)\} = \begin{Bmatrix} f_{m1} \\ f_{m2} \\ \vdots \\ f_{mn_c} \end{Bmatrix} = -[G_c]\{\xi_c(t)\} = [L_c]\{f_d(t)\} \tag{2.69}$$

where $[L_c] = [\Psi_c]^{-1}[B]$, $[\Psi_c]^{-1}$ is a $2n_c \times 2n$ matrix, which is constituted by first $2n_c$ row elements, and the above operation can effectively reduce the calculation. So, the final control force vector is expressed as:

$$\{f_d(t)\} = -[L_c]^+[G_c]\{\xi_c(t)\} \tag{2.70}$$

where $[L_c]^+$ is the pseudo-inverse matrix of $[L_c]$. Considering that the number of controllers is less than $2n_c$, and the $[L_c]^+$ is given by:

$$[L_c]^+ = \left([L_c]^T[L_c]\right)^{-1}[L_c]^T \tag{2.71}$$

In general case, the modal generalized coordinate is difficult to measure, so the $\{\xi_c(t)\}$ is expressed as the state vector of system:

$$\{\xi_c(t)\} = [\Psi_c]^{-1}\{Z(t)\} \tag{2.72}$$

2.5.2 Modal Control Based on Equation of Motion

In the field of civil engineering, it is easy to obtain the equation of motion of the structure, and the calculation of the real modes does not involve a complex operation. Thus the coupled equations of motion of the structure can be decoupled into n independent generalized coordinate equations by modal decomposition. Then n independent equations are rewritten as the state-space equation, and the optimal control force vector is calculated based on the Section 2.5.1.

The equation of motion of the structural system with a finite number n of degrees of freedom under dynamic loads can be obtained in Eq. (2.1), and the vector of displacement can be expressed as:

$$\{x(t)\} = [\Phi]\{q(t)\} \tag{2.73}$$

where $\{q(t)\}$ is the vector of generalized modal coordinate of the controlled structure. Substituting Eq. (2.73) into Eq. (2.1), n independent generalized coordinate equations can be achieved:

$$[M^*]\{\ddot{q}(t)\} + [C^*]\{\dot{q}(t)\} + [K^*]\{q(t)\} = \{P^*(t)\} + \{f_d^*(t)\} \tag{2.74}$$

where $[M^*] = [\Phi]^T[M][\Phi]$, $[C^*] = [\Phi]^T[C][\Phi]$, $[K^*] = [\Phi]^T[K][\Phi]$, $\{P^*(t)\} = [\Phi]^T\{P(t)\}$, $\{f_d^*(t)\} = [\Phi]^T[B]\{f_d(t)\} = [L]\{f_d(t)\}$. $[L]$ is a $n \times p$ matrix, p is the number of controllers. Generally, the structural dynamic responses are dominated by a small number of particular modes of structures, so the n_c modes of structure are controlled, and then the equation of motion of the controlled structure are expressed as:

$$[M_c^*]\{\ddot{q}(t)\} + [C_c^*]\{\dot{q}(t)\} + [K_c^*]\{q(t)\} = \{P_c^*(t)\} + \{f_c^*(t)\} \tag{2.75}$$

Rewriting the above equation of motion into the state-space equation, the control force can be calculated based on the steps in 2.5.1. The final control force vector is given by:

$$\{f_c^*(t)\} = -[G_c]\begin{Bmatrix} q_c(t) \\ \dot{q}_c(t) \end{Bmatrix} \tag{2.76}$$

Similarly, since the generalized mode, vector is difficult to measure, so the optimal control force vector $\{f_d(t)\}$ is expressed with the displacement and velocity of the controlled structure:

$$\{f_d(t)\} = [L_c]^{-1}\{f_c^*(t)\} = -[L_c]^+([G_{c1}][\Phi_c]^{-1}\{x(t)\} + [G_{c2}][\Phi_c]^{-1}\{\dot{x}(t)\}) \tag{2.77}$$

$$[L_c]^+ = \begin{cases} [L_c]^{-1} & n_c = p \\ [L_c^T L_c]^{-1}[L_c]^T & n_c > p \\ [L_c]^T[L_c L_c^T]^{-1} & n_c < p \end{cases} \tag{2.78}$$

The above content displays the IMSC algorithm is a relatively simple and effective structural vibration control method, so the IMSC is widely applied to vibration control for civil engineering structures. However, the IMSC algorithm is inapplicable for the control of flexible structures, because the effect of the modal control force on the uncontrolled modes should be considered.

2.6 H_∞ Feedback Control

H_∞ feedback control is an effective control method that makes up the blemish of classical modern control methods with good robustness [43]. Actually, H_∞ feedback control adopts the H_∞-norm of the sensitivity function (transfer function) of the controlled system as the performance index, which makes the error of the system in the worst-case interference minimum based on the H_∞-norm. The design of H_∞ feedback control can be conducted by solving some algebraic Riccati equations, and this common solution is based on the state-space theory.

2.6.1 H_∞ Norm

Many control problems can be converted to a standard H_∞ control problem, which is described as shown in Fig. 2.2, where G represents the controlled system, and K represents the designed controller. The state-space equations of the controlled closed-loop system can be expressed as follows:

$$\{\dot{Z}(t)\} = [A]\{Z(t)\} + [B_0]\{f_d(t)\} + [D_0]\{P(t)\} \tag{2.79}$$

$$\{Y_1(t)\} = [C_1]\{Z(t)\} + [B_1]\{f_d(t)\} \tag{2.80}$$

$$\{Y_2(t)\} = [C_2]\{Z(t)\} + [D_2]\{P(t)\} \tag{2.81}$$

where $\{Z(t)\}$ is the state vector of system, $\{f_d(t)\}$ is the control input vector, $\{P(t)\}$ is the external input vector, $\{Y_1(t)\}$ and $\{Y_2(t)\}$ are the control output vector and the measured output vector, respectively, which are similar to the $\{Y(t)\}$ in Eq. (2.5). The design of controller mainly involves two aspects: one is the controller should make the system asymptotically stable, the other is to make the H_∞-norm of the sensitivity function (transfer function) minimum. The controller can be designed as the following:

$$\{f_d(t)\} = -[G]\{Y_2(t)\} \tag{2.82}$$

where $[G]$ is a gain matrix.

To obtain the transfer function matrix of the system, the measured output vector $\{Y_2\}$ is substituted into Eq. (2.82), then Eq. (2.82) is rewritten as:

$$\{f_d(t)\} = -[G][C_2]\{Z(t)\} - [G][D_2]\{P(t)\} \tag{2.83}$$

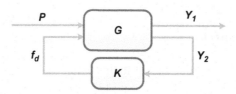

Figure 2.2
A standard control system.

Substituting Eq. (2.83) into Eq. (2.79) and Eq. (2.80), the state-space equations of the controlled system and the controller can be written as:

$$\{\dot{Z}(t)\} = [\tilde{A}]\{Z(t)\} + [\tilde{B}]\{P(t)\} \tag{2.84}$$

$$\{Y_1(t)\} = [\tilde{C}]\{Z(t)\} + [\tilde{D}]\{P(t)\} \tag{2.85}$$

where $[\tilde{A}] = [A] - [B_0][G][C_2]$, $[\tilde{B}] = -[B_0][G][D_2] + [D_0]$, $[\tilde{C}] = [C_1] - [B_1][G][C_2]$, $[\tilde{D}] = -[B_1][G][D_2]$. So, the transfer function matrix of system $[T_{Y_1 P}]$ can be given by:

$$[T_{Y_1 P}] = [\tilde{C}]([sI] - [\tilde{A}])^{-1}[\tilde{B}] + [\tilde{D}] \tag{2.86}$$

Thus the H_∞-norm of the transfer function matrix is defined as:

$$\|T_{Y_1 P}\|_\infty = \sup_{0 \le w < \infty} \sigma_{\max}[T_{Y_1 P}(jw)] \tag{2.87}$$

where $\sigma_{\max}[T_{Y_1 P}(jw)]$ is the maximum singular value of the transfer function. For the single-input single-output system, the maximum singular value of the transfer function is the maximum modules amplitude of transfer function.

2.6.2 H_∞ Feedback Control

The infimum of the H_∞-norm set of the transfer function of controllers that ensure the closed-loop system stable is expressed as:

$$\gamma_\infty^* = \inf\{\|T_{Y_1 P}\|_\infty \mid \text{the set of controllers that ensure system stable}\} \tag{2.88}$$

If a controller can ensure the closed-loop system stable and $\|T_{Y_1 P}\|_\infty < \gamma (\gamma > \gamma_\infty^*)$, the controller is defined as $H_\infty - \gamma$ suboptimal controller.

Generally, the measurements of all of the states of the controlled system are unrealizable, so the controller is designed based on the static output feedback approach. In any given case $\gamma > \gamma_\infty^*$, the controller can be designed as:

$$\{\dot{V}(t)\} = [A_c]\{V(t)\} + [B_c]\{Y_1(t)\} \tag{2.89}$$

$$\{f_d(t)\} = -[G]\{V(t)\} \tag{2.90}$$

where $\{V(t)\}$ is the state vector of the controller.

$$\begin{cases} [A_c] = [A] + \gamma^{-2}[D_0][D_0]^T[\overline{P}] - [B_0][G] - ([I] - \gamma^{-2}[Q][\overline{P}])^{-1}[\overline{K}] \\ \qquad ([C_2] + \gamma^{-2}[D_2][D_0]^T[\overline{P}]) \\ [B_c] = ([I] - \gamma^{-2}[Q][\overline{P}])^{-1}[\overline{K}] \\ [G] = ([B_1]^T[B_1])^{-1}([B_1]^T[C_1] + [B_0]^T[\overline{P}]) \\ [\overline{K}] = ([Q][C_2]^T + [D_0][D_2]^T)([D_2][D_2]^T)^{-1} \end{cases} \tag{2.91}$$

The matrixes $[\overline{P}]$ and $[Q]$ are the *semi-positive definite* solutions of the following Riccati matrix equations:

$$[A]^{T}[\overline{P}] + [\overline{P}][A] + [C_1]^{T}[C_1] + [\overline{P}][D_0][D_0]^{T}[\overline{P}]/\gamma^2 \\ - ([\overline{P}][B_0] + [C_1]^{T}[B_1])([B_1]^{T}[B_1])^{-1}([B_1]^{T}[C_1] + [B_0]^{T}[\overline{P}]) = 0 \quad (2.92)$$

$$[Q][A]^{T} + [A][Q] + [D_0][D_0]^{T} + [Q][C_1]^{T}[C_1][Q]/\gamma^2 - \\ ([Q][C_2]^{T} + [D_0][D_2]^{T})([D_2]^{T}[D_2])^{-1}([D_2]^{T}[D_0]^{T} + [C_2][Q]) = 0 \quad (2.93)$$

Sometimes, the measurements of all of the states of the controlled system are realizable, so the control is designed based on the full state feedback approach, and the controlled system and the controller are expressed as:

$$\{\dot{Z}(t)\} = [A]\{Z(t)\} + [B_0]\{f_d(t)\} + [D_0]\{P(t)\} \quad (2.94)$$

$$\{Y_1(t)\} = [C_1]\{Z(t)\} + [B_1]\{f_d(t)\} \quad (2.95)$$

$$\{Y_2(t)\} = \{Z(t)\} \quad (2.96)$$

The full state feedback controller is designed as:

$$\{f_d(t)\} = -[G]\{Z(t)\} = -([B_1]^{T}[B_1])^{-1}([B_1]^{T}[C_1] + [B_0]^{T}[\overline{P}])\{Z(t)\} \quad (2.97)$$

where $[\overline{P}]$ is the positive-definite solution of the following Riccati equation:

$$[A]^{T}[\overline{P}] + [\overline{P}][A] + [C_1]^{T}[C_1] + [\overline{P}][D_0][D_0]^{T}[\overline{P}]/\gamma^2 \\ - ([\overline{P}][B_0] + [C_1]^{T}[B_1])([B_1]^{T}[B_1])^{-1}([B_1]^{T}[C_1] + [B_0]^{T}[\overline{P}]) = 0 \quad (2.98)$$

The excellent robustness is an advantage of H_∞ feedback control in the intelligent control of civil engineering structures. In addition, the operation of H_∞ feedback control is relatively simple, and the only problem is the selection of the transfer function matrix of the controlled structure.

2.7 Sliding Mode Control

SMC method is widely applied into the control of variable structures, and the core of SMC is that the motion of the structure needs to tend to the defined sliding surface [44,45]. Meanwhile, the motion of the structure on the sliding surface defined is asymptotically stable. The SMC has an excellent robustness for the external excitation and the parameter perturbations of system, and many random disturbances and inevitable uncertainties exist in the process of structural vibration control. Therefore the SMC is more effective in control of both linear structures and nonlinear structures. Generally, the design of the SMC is determined by two aspects: the sliding surface and the controller.

2.7.1 Design of Sliding Surface

The equation of the linear time-invariant system can be expressed as Eq. (2.62), and the sliding surface can be designed as the linear combination of the state vectors of the system:

$$\{S\} = [\Theta]\{Z(t)\} \tag{2.99}$$

where $[\Theta]$ is a $p \times 2n$ matrix, which determines the stability and performance of the sliding mode, and the design of the sliding surface is to find the matrix $[\Theta]$. The $[\Theta]$ can be designed based on the LQR algorithm or the pole assignment algorithm, when the control is the full state feedback. If the static output feedback control is adopted, $[\Theta]$ can be designed based on pole assignment algorithm. Here the LQR algorithm is selected to design the sliding surface.

Firstly, the state vector of the system is conducted by a linear transformation:

$$\{y(t)\} = [T]\{Z(t)\} \tag{2.100}$$

where $[T]$ is a state-transition matrix, which can be expressed as:

$[T] = \begin{bmatrix} [I_{2n-p}] & -[B_1][B_2]^{-1} \\ [0] & [I_p] \end{bmatrix}$, $[T]^{-1} = \begin{bmatrix} [I_{2n-p}] & [B_1][B_2]^{-1} \\ [0] & [I_p] \end{bmatrix}$, $[B_0] = \begin{bmatrix} [B_1] \\ [B_2] \end{bmatrix}$; $[B_0]$ is the position matrix of installed controllers, which is rewritten as partition form, $(2n-p) \times p$ matrix $[B_1]$ and $p \times p$ matrix $[B_2]$. The former represents the position where controllers are not installed, the latter represents the position where p controllers are installed and is a nonsingular matrix. If the external load is ignored, the state equations of a system and sliding surface can be rewritten as:

$$\{\dot{y}(t)\} = [\tilde{A}]\{y(t)\} + [\tilde{B}]\{f_d(t)\} \tag{2.101}$$

$$[S] = [\tilde{\Theta}]\{y(t)\} \tag{2.102}$$

where $[\tilde{A}] = [T][A][T]^{-1}$, $[\tilde{B}] = [0 \quad B_2^T]^T$, $[\tilde{\Theta}] = [\Theta][T]^{-1}$. In order to achieve the feedback mechanism, the matrixes and vector in Eqs. (2.101) and (2.102) are rewritten as partition form: $\{y(t)\} = \{\{y_1(t)\}^T, \{y_2(t)\}^T\}^T$, $[\tilde{A}] = \begin{bmatrix} [\tilde{A}_{11}] & [\tilde{A}_{12}] \\ [\tilde{A}_{21}] & [\tilde{A}_{22}] \end{bmatrix}$, $[\tilde{\Theta}] = [[\tilde{\Theta}_1] \quad [\tilde{\Theta}_2]]$.

Then the state equations of system and sliding surface can be expressed as:

$$\{\dot{y}_1(t)\} = [\tilde{A}_{11}]\{y_1(t)\} + [\tilde{A}_{12}]\{y_2(t)\} \tag{2.103}$$

$$\{S\} = [\tilde{\Theta}_1]\{y_1(t)\} + [\tilde{\Theta}_2]\{y_2(t)\} = 0 \tag{2.104}$$

In order to simplify the derivation, the $[\tilde{\Theta}_2]$ is assumed as a unit matrix, then:

$$\{\dot{y}_1(t)\} = ([\tilde{A}_{11}] - [\tilde{A}_{12}][\tilde{\Theta}_1])\{y_1(t)\} \tag{2.105}$$

$$\{y_2(t)\} = -[\tilde{\Theta}_1]\{y_1(t)\} \tag{2.106}$$

Assuming that all of the states are measured, the performance index of the system is given by:

$$J = \int_{t_0}^{\infty} \{Z(t)\}^T [Q]\{Z(t)\} dt \tag{2.107}$$

where $[Q]$ is a positive definite weighting matrix. The performance index of the system can be expressed by $\{y(t)\}$ after a further transformation:

$$J = \int_{t_0}^{\infty} \{y(t)\}^T [\tilde{Q}]\{y(t)\} dt \tag{2.108}$$

where $[\tilde{Q}] = [T]^{-1T}[Q][T]^{-1} = \begin{bmatrix} [\tilde{Q}_{11}] & [\tilde{Q}_{12}] \\ [\tilde{Q}_{21}] & [\tilde{Q}_{22}] \end{bmatrix}$.

In order to minimize the above performance index and satisfy the constraint, $\{y_2(t)\}$ is expressed by the feedback state $\{y_1(t)\}$ based on the maximum principle:

$$\{y_2(t)\} = -0.5 [\tilde{Q}_{22}]^{-1}([\tilde{A}_{12}]^T[\tilde{P}] + 2[\tilde{Q}_{21}])\{y_1(t)\} \tag{2.109}$$

where $[\tilde{P}]$ is the solution of the following Riccati equation:

$$[\hat{A}]^T[\tilde{P}] + [\tilde{P}][\hat{A}] - 0.5[\tilde{P}][\tilde{A}_{12}][\tilde{Q}_{22}]^{-1}[\tilde{A}_{12}]^T[\tilde{P}] + \\ 2([\tilde{Q}_{11}] - [\tilde{Q}_{12}][\tilde{Q}_{22}]^{-1}[\tilde{Q}_{12}]^T) = 0 \tag{2.110}$$

where $[\hat{A}] = [\tilde{A}_{11}] - [\tilde{A}_{12}][\tilde{Q}_{22}]^{-1}[\tilde{Q}_{21}]$. So far, the sliding surface can be identified as

$$[\tilde{\Theta}_1] = -0.5[\tilde{Q}_{22}]^{-1}([\tilde{A}_{12}]^T[\tilde{P}] + 2[\tilde{Q}_{21}]) \tag{2.111}$$

$$[\Theta] = [\tilde{\Theta}][T] = [[\tilde{\Theta}_1]:[I_p]][T] \tag{2.112}$$

2.7.2 Design of Controller

The design of the controller is the second phase of SMC design, and the aim is to make the state of the system tend to the sliding surface and keep stable on the surface. Here the controller is designed based on the Lyapunov direct method.

Setting Lyapunov function as follows:

$$v = 0.5\{S\}^T\{S\} = 0.5\{Z(t)\}^T[\Theta]^T[\Theta]\{Z(t)\} \tag{2.113}$$

The sufficient condition to ensure the system performance stable on sliding surface is:

$$\dot{v} = \{S\}^T\{\dot{S}\} \leq 0 \tag{2.114}$$

Substituting Eq. (2.5) into Eq. (2.114), then

$$\dot{v} = \{S\}^T\{\dot{S}\} = \{S\}^T[\Theta]\{\dot{Z}(t)\} = \{S\}^T[\Theta]([A]\{Z(t)\} + [B_0]\{f_d(t)\} + [D_0]\{P(t)\})$$
$$= \{S\}^T[\Theta][B_0](\{f_d(t)\} + ([\Theta][B_0])^{-1}[\Theta]([A]\{Z(t)\} + [D_0]\{P(t)\}))$$
$$= \{\lambda\}(\{f_d(t)\} - \{G_s\}) = \sum_{i=1}^{p} \lambda_i(f_{di} - G_{is}) \tag{2.115}$$

According to the above sufficient condition in Eq. (2.114), the control force derived from ith controller can be designed as

$$f_{di}^*(t) = \begin{cases} G_{is} - W_i \,\text{sgn}(\lambda_i); & \text{if } \sum_{i=1}^{p} -\lambda_i G_{is} \geq 0 \\ 0; & \text{if } \sum_{i=1}^{p} -\lambda_i G_{is} < 0 \end{cases} \tag{2.116}$$

where W_i is a given smaller constant, $W_i = (0.0001 \sim 0.0005)f_{d,\max}$. Considering the maximum driving force of the actuator, the final control force is given by:

$$f_{di}(t) = \begin{cases} f_{di}^*(t); & \text{if } |f_{di}^*| < f_{d,\max} \\ f_{d,\max}\,\text{sgn}(f_{di}^*(t)); & \text{others} \end{cases} \tag{2.117}$$

The SMC is conducted by the above operation in civil engineering structures, in which the sliding surface can be defined as the linear combination of the state vectors of the controlled structure, and the controller can be designed through the Lyapunov direct method or the other control laws. Generally, the SMC is applied to the control that the parameters of the controlled system constantly change, such as the semi-active variable stiffness and variable damping controls.

2.8 Optimal Polynomial Control

The canonical theory in the stochastic optimal control of structures is subjected to an open challenge to the practical nonstationary, non-Gaussian white noise excitation system, such as earthquake ground motions, strong winds and sea waves which are usually encountered in civil engineering structures. For replying this challenge, a physical approach to structural stochastic optimal controls, i.e., OPC, has been proposed. Compared with nonlinear optimal control methods, such as instantaneous optimal control, SMC, generalized optimal control, etc. the OPC is considered as one of the preferred methods for nonlinear systems [46].

2.8.1 Basic Principle

The physical stochastic optimal control is implemented by solving a collection of deterministic dynamic equations corresponding to the representative realizations with preassigned weights [47]. Hence it is spontaneous to develop a nonlinear stochastic optimal control strategy by integrating a physical stochastic control scheme and a deterministic nonlinear optimal control theory. According to the original physical scheme, the nonlinear stochastic optimal control strategy can be performed and optimized as follows [48]. Firstly, for each representative realization of stochastic parameters, the minimization of a cost function is executed to build a functional mapping from the set of parameters of control policy to the set of control gains. Then the specified parameters of control policy are obtained by minimizing a performance function related to the objective structural performance, as well as the corresponding control gain. The content mentioned above is the fundamental theory of stochastic optimal polynomial.

In Section 2.1, the motion equation of the structural system with a finite number n degrees of freedom under dynamic loads has been given as Eq. (2.1), which can be expressed as the following after mathematical simplification

$$[M]\{\ddot{x}(t)\} + \{f[x(t), \dot{x}(t)]\} = [B]\{f_d(t)\} + [D]\{F(\varpi, t)\},$$

$$x(t_0) = x_0, \dot{x}(t_0) = \dot{x}_0 \tag{2.118}$$

where $\{x(t)\} = \{x(\Theta, t)\}$ is an n-dimensional displacement vector; $\{f_d(t)\} = \{f_d(\Theta, t)\}$ is an r-dimensional control force vector; $[M]$ is a $n \times n$ mass matrix; $\{f[\bullet]\}$ is a n-dimensional vector denoting nonlinear internal forces, including a nonlinear damping force and a nonlinear restoring force; $\{F[\bullet]\}$ is a p-dimensional random excitation vector, in which ϖ represents a point in the probability space, i.e., an embedded basic random event characterizing the randomness inherent in the external excitation. $\{\Theta\} = \{\Theta(\varpi)\}$ denotes a random parameter vector mapped from ϖ, which implicitly underlies the state and control force vectors. $[B]$ is a $n \times r$ matrix denoting the location of controllers; $[D]$ is a $n \times p$ matrix denoting the location of excitations. x_0 and $\dot{x}_0$ are the initial displacement and initial velocity, respectively.

The classical OPC theory was proposed on the basis of Hamilton−Jacobi theoretical framework and the optimal principle [49], which is essentially the extended formulation of the LQR control in state space. In order to express Eq. (2.118) as the form of its state equation, the nonlinear internal force $\{f[\bullet]\}$ needs to be expanded. It is usually expressed as the following Maclaurin series:

$$\{f[x(t), \dot{x}(t)]\} = x\left\{\frac{\partial f[x,\dot{x}]}{\partial x} + \frac{1}{2!}x\frac{\partial^2 f[x,\dot{x}]}{\partial x^2} + \cdots + \frac{1}{m!}x^{m-1}\frac{\partial^m f[x,\dot{x}]}{\partial x^m}\right\}\bigg|_{x=0,\dot{x}=0}$$

$$+ \dot{x}\left\{\frac{\partial f[x,\dot{x}]}{\partial \dot{x}} + \frac{1}{2!}\dot{x}\frac{\partial^2 f[x,\dot{x}]}{\partial \dot{x}^2} + \cdots + \frac{1}{m!}\dot{x}^{m-1}\frac{\partial^m f[x,\dot{x}]}{\partial \dot{x}^m}\right\}\bigg|_{x=0,\dot{x}=0} \tag{2.119}$$

where m is the highest order of the Maclaurin series, and is equal to the order of the nonlinear internal force, indicating that the terms of series with $(m + 1)$ and the higher-order parts are zero. Meanwhile, the zero order terms and the $x^i \dot{x}^j$ cross terms which make little contribution to the nonlinear internal force of engineering structure are omitted.

The state-space equation has been given as Eq. (2.5), with the initial state $\{Z(t_0)\} = z_0$; $\{P(t)\}$ is same as $\{F(\varpi, t)\}$; and $[A]$ can be expressed as the following after mathematical deformation:

$$[A] = \begin{bmatrix} [0] & [I] \\ -[M]^{-1} \sum_{i=1}^{m} \frac{x^{i-1}}{i!} \frac{\partial^i f[x,\dot{x}]}{\partial x^i} \bigg|_{x=0, \dot{x}=0} & -[M]^{-1} \sum_{i=1}^{m} \frac{\dot{x}^{i-1}}{i!} \frac{\partial^i f[x,\dot{x}]}{\partial \dot{x}^i} \bigg|_{x=0, \dot{x}=0} \end{bmatrix} \quad (2.120)$$

A polynomial cost function, in stochastic case with $\{\Theta\}$ [50], is given by

$$J_1(\{Z\}, \{f_d(t)\}, \{\Theta\}, t) = S(\{Z(t_f)\}, t_f) + \frac{1}{2} \int_{t_0}^{t_f} [\{Z(t)\}^T [Q_Z] \{Z(t)\} \\ + \{f_d(t)\}^T [R_f] \{f_d(t)\} + \{h(Z, t)\}] dt \quad (2.121)$$

where $S(\{Z(t_f)\}, t_f)$ is the terminal cost; $\{Z(t_f)\}$ is the terminal state; t_0 and t_f are the start and terminal time, respectively; $[Q_Z]$ is a $2n \times 2n$ positive semi-definite matrix denoting state weighting; $[R_f]$ is a $r \times r$ positive-definite matrix denoting control weighting; $\{h(Z, t)\}$ is the higher-order term of the cost function whose order is higher than the quadratic term.

The minimization of the polynomial cost equation will result in the celebrated Hamilton−Jacobi−Bellman equation [51]:

$$\frac{\partial \{V(Z, t)\}}{\partial t} = - \min_{f_d} [\{H(Z, f_d(t))\}, \{V'(Z, t)\}, \{\Theta\}, t)] \quad (2.122)$$

where the optimal cost function $\{V(Z, t)\}$ satisfies all the properties of a Lyapunov function, considered as [51]

$$\{V(Z, t)\} = \frac{1}{2} \{Z(t)\}^T [\tilde{P}(t)] \{Z(t)\} + \{g(Z, t)\} \quad (2.123)$$

where $[P(t)\tilde{\ }]$ is a $2n \times 2n$ Riccati matrix [48]; $\{g(Z, t)\}$ is a positive-definite multinomial in $\{Z(t)\}$. The Hamiltonian function $\{H(\bullet)\}$ is defined by [50]

$$\{H(Z, f_d(t), V'(Z, t), \Theta, t)\} = \frac{1}{2} [\{Z(t)\}^T [Q_Z] \{Z(t)\} + \{f_d(t)\}^T [R_f] \{f_d(t)\} + \{h(Z, t)\}] \\ + \{V'(Z, t)\}^T ([A] \{Z(t)\} + [B_0] \{f_d(t)\}) \quad (2.124)$$

The necessary condition for the minimization of the right-hand side of Eq. (2.122) is

$$\frac{\partial \{H(Z, f_d(t), V'(Z,t), \Theta, t)\}}{\partial f_d} = 0 \tag{2.125}$$

The optimal nonlinear controller is then given by

$$\{f_d(t)\} = -[R_f]^{-1}[B_0]^T[\widetilde{P(t)}]\{Z(t)\} - [R_f]^{-1}[B_0]^T\{g'(Z,t)\} \tag{2.126}$$

To express the controller Eq. (2.126) as an explicit function, $\{g(Z,t)\}$ is chosen as the following [50]

$$\{g(Z,t)\} = \sum_{i=2}^{k} \frac{1}{i}[\{Z(t)\}^T[M_i(t)]\{Z(t)\}]^i \tag{2.127}$$

where $[M_i(t)]$, $i = 2, 3, \ldots, k$ are $2n \times 2n$ Lyapunov matrices [49].

Both $[\widetilde{P}(t)]$ and $[M_i(t)]$ are related to the gradient matrix $[A]$. Therefore the gain matrices of the polynomial controller cannot be calculated off-line. An approximate solution to the gain matrices is obtained by linearizing the gradient matrix $[A]$ at the initial equilibrium point z_0 [50], i.e., replacing $[A]$ into $[A_0] = [A]|_{z_0}$. Furthermore, the Riccati matrix $[P(t)]$ and the Lyapunov matrices $[M_i(t)]$, for a class of optimal control system with time infinite, could be approximately evaluated as constant matrices ($[\widetilde{P}]$ and $[M_i]$) by solving the following algebraic Riccati and Lyapunov equations, i.e., the steady-state Riccati and Lyapunov matrix equations, respectively

$$[\widetilde{P}][A_0] + [A_0]^T[\widetilde{P}] - [\widetilde{P}][B_0][R_f]^{-1}[B_0]^T[\widetilde{P}] + [Q_z] = 0 \tag{2.128}$$

$$[M_i]([A_0] - [B_0][R_f]^{-1}[B_0]^T[\widetilde{P}]) + ([A_0] - [B_0][R_f]^{-1}[B_0]^T[\widetilde{P}])^T[M_i] + [Q_{z,i}] = 0 \tag{2.129}$$

where $i = 2, 3, \ldots, k$.

Eqs. (2.128) and (2.129) can be solved using any well-known numerical algorithms or by virtue of computing software, e.g., MATLAB.

Hence an optimal polynomial controller can be obtained analytically in the form

$$\{f_d(t)\} = -[R_f]^{-1}[B_0]^T[\widetilde{P}]\{Z(t)\} - [R_f]^{-1}[B_0]^T \sum_{i=2}^{k}[\{Z^T(t)\}[M_i(t)]\{Z(t)\}]^{i-1}[M_i]\{Z(t)\} \tag{2.130}$$

Eq. (2.130) indicates that the polynomial controller consists of two components, the linear term which is the first order and the nonlinear term which is the odd higher order.

2.8.2 Applications

As an extended conventional LQR, OPC provides more flexibility in the control design and further enhances system performance. Under strong earthquakes, the main objective of intelligent control is to reduce the peak (maximum) response of the structure in order to minimize the damage. For both linear and nonlinear or hysteretic structures, it has been shown that the polynomial controller is more effective than the classical linear controller in suppressing the peak response, due to its ability to provide large control force under strong earthquakes [50, 52].

At present, the optimal polynomial controller is widely applied in the control of highly nonlinear or inelastic hybrid protective systems, such as the base-isolated building using lead-core rubber bearings and the fixed-base yielded building. The performances of the optimal polynomial controller with respect to various control objectives are investigated by means of numerical simulations [50]. The results have shown obvious advantages of the optimal polynomial controller in the field of structural vibration control.

2.9 Fuzzy Control

In the field of traditional control, whether the dynamic model of the control system is accurate or not directly dominates the calculation results. The more detailed the dynamic information of the control system is, the easier to achieve the purpose of precise control. However, for complex systems, it is often very difficult to describe the dynamic characteristics of the systems accurately due to too many variables. Thus engineers always use various methods to simplify the dynamic characteristics of the systems to achieve the control purpose, while the result is still not ideal. In other words, the traditional control theory can deal with the relatively simple systems very well, but it will be unideal or powerless for the systems which are very complicated or difficult to be described accurately. The control strategies introduced in the previous sections belong to the traditional control strategies, and they can achieve good control effects in their applicable conditions. But civil engineering structures are complicated systems with the characteristics of strong nonlinear, time-varying and time-delay, so it is difficult to establish their accurate models. Fuzzy control is the combination of fuzzy logical theory and the control technology, which is essentially a kind of nonlinear control and belongs to the category of intelligent control. The fuzzy control algorithm does not need the accurate mathematical model of controlled systems, and can achieve real-time control, according to the input and output data of the actual system and the expert's knowledge or operating experience. The fuzzy control method will be introduced in this section in detail.

2.9.1 Basic Principle

The core of fuzzy control method is the fuzzy controller. The fuzzy controller mainly includes three parts: fuzzification, fuzzy reasoning, and defuzzification, and its basic structure is shown in Fig. 2.3.

A fuzzification of mathematical concepts is based on a generalization of these concepts from characteristic functions to membership functions. That is, fuzzification of accurate input values means the external excitation acquired by computer sampling is translated into the membership functions. The purpose of fuzzification is to change the variable type such that it can be accepted and operated by the knowledge base.

The formation and reasoning of fuzzy rules are based on the control rules which are designed by the experienced operator or expert. A fuzzy output set is obtained by fuzzy reasoning, namely a new fuzzy membership function. The purpose of fuzzy reasoning is to adapt the control rules, determine the fitness of each control rules, and then the outputs are gained by fuzzy rules weighted.

The output value can be obtained by defuzzification according to the output fuzzy membership function, and different methods can be used to find a representative accurate value as the control value. The purpose of defuzzification is to obtain the accurate output value, which can be applied to the actuator to realize control.

2.9.2 Design of Fuzzy Controller

The design of fuzzy controller mainly consists of the following parts.

2.9.2.1 Determination of the basic domain

After the input and output variables of the fuzzy controller are chosen, the basic domain should be determined subsequently. For the determination of basic domain of input variables, it should be determined according to the characteristics of the whole controlled systems. The choice of the basic domain is very important. If the basic domain is too small, the normal data may exceed the threshold and will influence the performance of the system, such as, oscillation, amplification, and even divergence. On the contrary, if the field is too large the system responses will be slow, the output of the system cannot converge quickly

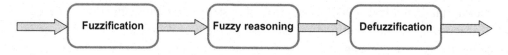

Figure 2.3
Basic structure of the fuzzy controller.

to the expected value. At the same time, the basic domain of the output value is also very important. If the domain is too large, the invalid area will be enlarged and lead to the oscillation of the system responses. If the domain is too small, the time of the system response will be extended and maybe cannot meet expectations.

2.9.2.2 Fuzzification of the accurate value

For the fuzzy controller, the input and output values must be fuzzy values. However, the input and output values of the actual controlled systems all are accurate values. Thus the fuzzification method should be employed to transform the accurate value to the fuzzy values. This process can be divided into two steps, the first step is to determine the fuzzy domain, and the second step is fuzzification.

The following formula can be adopted to transform the basic domain into the fuzzy fiedomain,

$$y = \frac{2n}{b-a}\left[x - \frac{a+b}{2}\right] \quad (2.131)$$

where x and y are accurate variable and fuzzy variable, respectively; $[a, b]$ is the basic domain of x; $[-n, +n]$ is the fuzzy domain of y; and n is a positive integer which is larger than 2. In fact, Mamdani method often is used, i.e., $n = 6$. In addition, then variable x in the domain $[a, b]$ also can be transformed into the asymmetric domain $[-n, +m]$, where n and m are both positive integers which are larger than 2. The transformation formula is as follows:

$$y = \frac{m+n}{b-a}\left[x - \frac{a+b}{2}\right] \quad (2.132)$$

Fuzzification is to divide the continuous quantity in fuzzy domain into several levels, according to the requirement, each level can be regarded as a fuzzy variable and corresponds to a fuzzy subset or a membership function. Usually, the fuzzy domain can be divided into the following the fuzzy subsets described as language variable: [NB (negative big), NM (negative middle), NS (negative small), Z (zero), PS (positive small), PM (positive middle), PB (positive big)], respectively.

2.9.2.3 Parameter selection

The parameters of fuzzy controller mainly include quantization factor and proportional factor. In order to transform the accurate value from basic domain into the corresponding fuzzy domain, the quantization factors should be employed, e.g., error quantization factor K_e and the error rate quantization $K_{\dot{e}}$. In addition, the control value deduced from the fuzzy control algorithm cannot be used to control systems directly, and it should be transformed into the basic domain. Thus the proportion factor K_u should be employed.

The quantization factors and the proportion factors dominate the performance of fuzzy controller directly. Taking the dynamic property of the system, e.g., the larger the quantization factor, the larger the overshoot and the transition process of the system. The reason is that the basic domain will be narrow with the increasing of K_e, the control effect of the error variance will increase, which will lead to the overshoot phenomenon and the extension of the transition process. If the error rate quantization $K_{\dot{e}}$ is chosen comparatively larger, the system overshoot will be small and the system response time will be extended. As for the static performance of the system, the increase of quantization factor K_e and $K_{\dot{e}}$ can reduce the steady-state error and the error rate, but the steady-state error cannot be eliminated. The small output proportion factor K_u will extend the dynamic process, while large K_u will lead to the oscillation of the system.

2.9.2.4 Selection of the membership function

The membership function is the characteristic function of the fuzzy set, and it is the basis to solve practical problem using the fuzzy theory. Thus it is very important to select an appropriate membership function, and the following principles should be considered.

1. The shape of the membership function has direct influence on the system stability, which is usually restricted to the convex fuzzy set. The functions with high resolution are usually adopted in the areas with small error to increase the sensitivity; while the functions with comparatively low resolution are usually used in the large error areas to guarantee the satisfying robustness of the system.
2. The membership function should comply with the semantic sequence to avoid inappropriate overlap, and the overlap degree directly influences the performance of the system.
3. The distributions of the membership functions are always symmetric and balanced.

2.9.2.5 Determination of the rule base

The essence of the fuzzy control rules is summarizing the experience of the operator or the knowledge of the experts to form the control rules, which consists of the rule base.

The fuzzy rules should have the characteristics of ε completeness, consistency, and interaction. The ε completeness characteristic means that a rule can be always found for each input state, which can match the output in a certain larger than $\varepsilon(\varepsilon \in [0, 1])$. The consistency characteristic means that the formation of the fuzzy rule is based on the knowledge of the experts and the experience of the operator, and the rule can match different performance standard. For a complete rule base, if the input is x_i, and the expectable output is $u_i(i = 1, 2\cdots, n)$, the actual control force is usually not equal to u_j. In other word, taking R to represent the fuzzy relation, then the rules will have the interaction characteristics if they fit the following equation $\underset{1 \leq i \leq n}{\exists} \underset{u \in U}{\exists} (x_i \circ R)(u) \neq u_i$.

2.9.2.6 Defuzzification

The results of fuzzy reasoning are fuzzy set or membership functions, and they reflect the combination of different control languages. However, the controlled objective can only accept one control accurate quantity, thus one control accurate quantity should be chosen from the output fuzzy set. In other word, the method of defuzzification is to deduce a mapping from fuzzy set to the ordinary number set. The commonly used method are the coefficient weighted mean method, the gravity method, and the maximum membership degree method. These methods will be introduced as follows:

1. The coefficient weighted mean method
   The output quantity of this method can be expressed as,

$$u = \sum k_i \cdot x_i / \sum k_i \tag{2.133}$$

where u is the output value of solving fuzzy, x_i is the fuzzy input, k_i is the coefficient; the selection of k_i should depend on the actual condition, and different coefficients will result in different response characteristics.

2. The gravity method
   The gravity method means, that the gravity of the area enveloped by the fuzzy membership function curve and the output value is chosen as the abscissa. The following equation can be employed,

$$u = \frac{\int x \mu_N(x) dx}{\int \mu_N(x) dx} \tag{2.134}$$

where x is the fuzzy input, and $\mu_N(x)$ is the membership degree function. In the actual calculation process, the following simple equation is adopted,

$$u = \sum x_i \cdot \mu_N(x_i) / \sum \mu_N(x_i) \tag{2.135}$$

   In addition, the gravity method is the special case of the coefficient weighted mean method, and they are same when $k_i = \mu_N(x_i)$.
3. The maximum membership degree method

The maximum membership degree method is one of the simplest methods, and only the element with the maximum membership degree is chosen as the output value. What is more, the curve of the membership function should be convex (unimodal curve). If the curve is flat trapezoid, the element with a maximum membership degree is larger than 1 and the average value should be adopted.

2.10 Neural Network Control

Neural network control method is a kind of mathematical model of the distributed parallel algorithm of information processing, which can imitate animal neural network behavior characteristics. This kind of network depends on the complexity of the system to achieve the purpose of processing information by adjusting the internal relations between a large number of nodes connected, and has the ability of self-learning, self-adaptive and self-organization. Neural networks have been used to solve problems in almost all spheres of science and technology. For civil engineering structures having the characteristics of strong nonlinear, time-varying and time-delay, neural network control algorithm can be used to identify the models of the controlled civil engineering structures, predict the responses of the structures, and so on. This section will focus on the basic theory of the neural network control method.

2.10.1 Basic Principle

The neural network is proposed by studying the human information according to the modern neurobiology and understanding science. It can be seen as a network which is composed of a large number of interconnected neurons (processing unit). The neural work has the characteristics of strong adaptability, learning ability, nonlinear mapping ability, robustness, and fault tolerance.

The neurons can transmit information, and their models can be classified as three parts: threshold unit, linear unit, and nonlinear unit. Once the model of neuron is determined, the performance and ability of a neural network will mainly depend on the topological structure and the learning method. The topological structures of neural networks are usually divided as

1. *Feedforward network.* The neurons of the network exist layer by layer, and each neuron is only connected to other neurons in the former layer. The top layer is the output layer, and the number of the hidden layer can be one or more. The feedforward network is widely used in the practical application such as sensors.
2. *Feedback network.* The network itself is a forward model. The difference between the feedforward and the feedback model is that the feedback network has a feedback loop.
3. *Self-organizing network.* The lateral inhibition and exciting mechanism of the neurons between the same layer can be realized through the interconnection of the neurons, thus the neurons can be classified into several categories, and each category can action as an integral.
4. *Interconnection network.* The interconnection network includes two categories: local interconnection and global interconnection. The neuron is connected to all the other neurons, while some of the neurons are not connected in the local interconnection network. The Hopfield network and the Boltzmann network belong to the interconnection network.

2.10.2 Learning Method

Once the topological structure of the neural network is determined, the learning method is indispensable to attribute the network with an intelligent characteristic, which is also the core problem in intelligent control using the neural network method. Through learning, neural networks can grasp the characteristics of intelligent or nonlinear structures. The learning method is actually the adjusting method of the weight of network connection. There are three major learning paradigms, each corresponding to a particular abstract learning task. These are supervised learning, unsupervised learning, and reinforcement learning. When the supervised learning model is used, the output of the network and the desired output signal (i.e., supervised signals) are compared, and then according to the difference the network's weights will be adjusted, finally differences will be smaller. When the unsupervised learning model is used, after the input signals are sent into the network, according to the present rules (such as competition rules), the network's weights will be adjusted automatically, and the network will have pattern classification function eventually. The reinforcement learning method is a kind of learning method between the above two. There are several basic learning methods such as Hebb learning the rules, Delta (δ) learning rules, Probability learning rules, Competition learning rules and Levenberg–Marquart algorithm, and so on. As an example of learning methods, the commonly used Levenberg–Marquart algorithm will be introduced in the following.

Levenberg–Marquart algorithm belongs to the supervised learning model. When the output value $\hat{y}$ is not equal with the expected output y, the error signal will propagate from the output end in the reverse direction of the net, and the weight and threshold values are always modified to make the output value close to the desired output. When sample $p(p = 1, 2, \ldots, P)$ is weight adjusted, it will be delivered to the other sample models for learning train until P times are completed.

The quadratic error function of the input and output modes of each sample P is defined as,

$$E_p = \frac{1}{2}\sum_{l=1}^{L}[y_p(l)-\hat{y}_p(l)]^2 = \frac{1}{2}\sum_{l=1}^{L} e_l^2 \qquad (2.136)$$

Thus the error function of the whole system is,

$$E = \frac{1}{2}\sum_{p=1}^{P}\sum_{l=1}^{L}[y_p(l)-\hat{y}_p(l)]^2 = \sum_{p=1}^{P} E_p \qquad (2.137)$$

For simplicity, the subscript p is omitted in the following equations, and the PIF is chosen as the error function of the system, thus

$$E = \frac{1}{2}\sum_{l=1}^{L}[y(l)-\hat{y}(l)]^2 = \frac{1}{2}\sum_{l=1}^{L} e_l^2 \qquad (2.138)$$

where L is the number of the output neuron, e_l is the error of the lth output neuron.

The weighting adjustment should be proceeded in the reverse direction of the gradient of function E to make the output close to the desired value as far as possible. When Levenberg–Marquardt algorithm is employed in the training process of feedback neural network, the weighting matrix can be expressed as [53]

$$[w^{i+1}] = [w^i] - \left[\frac{\partial^2 E}{\partial w^{i2}} + \mu[I]\right]^{-1}\left[\frac{\partial E}{\partial w^i}\right] \quad (2.139)$$

where i is the training steps, $[\partial E/\partial w^i]$ is the descent gradient of the performance function E with respect to the weighting matrix $[w^i]$, $\mu(\mu \geq 0)$ is the control factor, $[I]$ is the unit matrix, and the Jacobi matrix of the weighting value can be obtained by the Taylor series expansion of the error vector $[e]([e] = [e_1, e_2, \ldots, e_L]^T)$,

$$[J^i] = \begin{bmatrix} \frac{\partial e_1}{\partial w^i_1} & \frac{\partial e_1}{\partial w^i_2} & \cdots & \frac{\partial e_1}{\partial w^i_B} \\ \frac{\partial e_2}{\partial w^i_1} & \frac{\partial e_2}{\partial w^i_2} & \cdots & \frac{\partial e_2}{\partial w^i_B} \\ \vdots & \vdots & \ddots & \vdots \\ \frac{\partial e_L}{\partial w^i_1} & \frac{\partial e_L}{\partial w^i_2} & \cdots & \frac{\partial e_L}{\partial w^i_B} \end{bmatrix}_{L \times B} \quad (2.140)$$

where L is the number of output neuron, and B is number of weighting value. Combining Eqs. (2.139) and (2.140) can obtain the following equation,

$$[w^{i+1}] = [w^i] - [[J^i]^T[J^i] + \mu[I]]^{-1}[J^i]^T[e] \quad (2.141)$$

Eq. (2.141) is the key equation of Levenberg–Marquardt algorithm. The application of intelligent control on civil engineering structures by using Levenberg–Marquardt algorithm and neural network can be seen in the study [54].

2.11 Particle Swarm Optimization Control

For vibration control, it is always the multiobjective optimization control. Firstly, the main control objective is to reduce the structural displacement responses to meet the request of anti-earthquake for civil engineering structures and the standard of civil engineering structures, and this objective is to ensure the safety of civil engineering structures. Secondly, the other control objective is to reduce structural acceleration responses, and this objective will influence the safety of furniture and decoration in civil engineering structures. The particle swarm optimization algorithm belongs to swarm intelligence, and it is a kind of probability search algorithm with the following advantages: (1) there are not centralized control constraints. That is, individual fault cannot affect the whole problem solving, which will ensure the systems having stronger robustness. (2) The method of the nondirect communication is used to ensure the scalability of the systems. (3) The parallel distributed algorithm model is used, which can make the best of multiple processors. (4) There are no special requirements for the continuity of the problem definition. (5) The algorithm is simple, and it is easy to implement. Especially, the multipoint parallel search feature of the PSO algorithm makes it not only suitable for single objective optimization control, and also it can be applied to multiobjective optimization control.

2.11.1 Basic Principle

In 1995, based on the idea of a bird flock foraging, James Kennedy, American social psychologist, and Russell Eberhart, electrical engineer, proposed the PSO algorithm [55−59]. That is, the birds in the bird flock are abstracted as "particles" without mass and volume, and these "particles" will mutually collaborate and share information to update their current movement velocity and direction according to their own and swarm's best historical movement state information. This kind of method can well coordinate the movement relationship between the particles themselves and swarm, and look for the optimal solution in the complex solution space. The following, the basic PSO algorithm and the improved PSO algorithms will be described.

2.11.1.1 The basic PSO algorithm

Assuming that,

$\mathbf{x}_i = (x_{i1}, x_{i2}, \cdots x_{in})$ is the current position of the ith particle;

$\mathbf{v}_i = (v_{i1}, v_{i2}, \cdots v_{in})$ is the current velocity of the ith particle;

$\mathbf{p}_i = (p_{i1}, p_{i2}, \cdots p_{in})$ is the optimal position once experienced of the ith particle, and the optimal position stands for the optimal adaptive value. For the minimization problem, the smaller the objective value is, the better adaptive value is;

$\mathbf{p}_g = (p_{g1}, p_{g2}, \cdots p_{gn})$ is the optimal location of all particles of the swarm that experienced.

The evolution equation of the basic PSO equation can be described as:

$$v_{ij}(t+1) = v_{ij}(t) + c_1 * rand_1()*(p_{ij}(t) - x_{ij}(t)) + c_2 * rand_2()*(p_{gj}(t) - x_{ij}(t)) \quad (2.142)$$

$$x_{ij}(t+1) = x_{ij}(t) + v_{ij}(t+1) \quad (2.143)$$

where subscript j denotes the j dimension of the particles, subscript i denotes the ith particle of the swarm, t is the generation, c_1 is the cognitive learning coefficient, c_2 is the social learning coefficient, and $rand_1()$ and $rand_2()$ are the dependent random value within the range of [0, 1].

There are three parts in Eq. (2.142). The first part is the former velocity of the particle, which makes the algorithm have the capability of global research. The second part (cognition part) means the process of the particle's absorption of their own experience. The third part (social part) means the process of the particle absorption of the other particle. The model without the second part is called the social-only model, which converges quickly and may stick in the local optimal value for the complex problem. The model without the third part is called the cognition-only model, which cannot obtain the optimal value due to no contact between the individuals.

2.11.1.2 Improved PSO algorithm

As discussed above, the first part of Eq. (2.142) is to guarantee the global converging ability of the PSO algorithm, and the last two parts are to guarantee the local converging ability. Different balance relations exist between the global searching ability and the local searching ability for different problems, and balance relations should also be modified at any moment, but the basic PSO algorithm does not have this ability. Thus Shi and Eberhart [60] proposed the PSO algorithm with inertial weight, and the evolution equation is

$$v_{ij}(t+1) = \omega^* v_{ij}(t) + c_1^* rand_1()^*(p_{ij}(t) - x_{ij}(t)) + c_2^* rand_2()^*(p_{gj}(t) - x_{ij}(t)) \quad (2.144)$$

$$x_{ij}(t+1) = x_{ij}(t) + v_{ij}(t+1) \quad (2.145)$$

where ω is the inertia weight, which determines the influence degree of the former velocity on the current velocity, thus it can balance the effect of the global and local searching ability. ω can be a positive constant, positive linear or nonlinear function with respect to time.

In order to control the flight speed to balance the effect of the global and local searching ability, Clerc [61] proposed the PSO algorithm with constriction factor in 1999, then Eberhart and Shi [62] simplified the expression for practical use in 2000. The velocity evolution equation can be written as:

$$v_{ij}(t+1) = k^*(v_{ij}(t) + c_1^* rand_1()^*(p_{ij}(t) - x_{ij}(t)) + c_2^* rand_2()^*(p_{gj}(t) - x_{ij}(t))) \quad (2.146)$$

$$k = \frac{2}{|2 - \varphi - \sqrt{\varphi^2 - 4\varphi}|} \quad (2.147)$$

where k is the constriction factor, $\varphi = c_1 + c_2$ and $\varphi > 4$.

In fact, the velocity evolution equation of the PSO algorithm with constriction factor at $c_1 = c_2 = 2.05, \varphi = 4.1, k = 0.7298$, is same as the particle swarm algorithm with inertial weight at $c_1 = c_2 = 1.4962, \omega = 0.7298$.

In order to overcome premature convergence, Riget and Vesterstrøm [63] proposed the ARPSO algorithm to improve the performance of the algorithm based on the PSO, and used the diversity to measure and control the swarm characteristics to avoid premature convergence. The "attractive" and "repulsive" operators are employed to increase the efficiency. The velocity evolution equation can be written as,

$$v_{ij}(t+1) = v_{ij}(t) + dir^*(c_1^*rand_1()^*(p_{ij}(t) - x_{ij}(t)) + c_2^*rand_2()^*(p_{gj}(t) - x_{ij}(t))) \quad (2.148)$$

$$dir = \begin{cases} -1, & dir > 0 \& diversity < d_{low} \\ 1, & dir < 0 \& diversity > d_{high} \end{cases} \quad (2.149)$$

The diversity function of the swarm is,

$$diversity(S) = \frac{1}{|S| \cdot |L|} \cdot \sum_{i=1}^{|S|} \sqrt{\sum_{j=1}^{N} (p_{ij} - \bar{p}_j)^2} \quad (2.150)$$

where d_{low} is the upper limit of the target of the swarm diversity, d_{high} is the upper limit of the target of the swarm diversity, S is the number of the particles of the swarm, N is the dimension, p_{ij} is the jth component of the ith particle, and $\bar{p}_j$ is the average value of the jth components of all the particles.

During the operation of the algorithm, if the swarm diversity function satisfies $diversity(S) < d_{low}$, then $dir = -1$ and the population will keep away from the optimal position; if the swarm diversity function increases and exceeds d_{high}, then $dir = 1$ and the population will move close to the optimal position.

2.11.2 Design Procedure of the PSO Algorithm

For the different PSO algorithms, design steps are usually as follows:

1. To determine the problem representation scheme (coding scheme)
   When the PSO algorithm is used to solve problems, firstly, solutions of the problems should be mapped from the solution space to the representation space with a certain structure, i.e., the solutions of the problems are expressed by specific code series. According to the feature of problems, appropriate coding method is selected, which will affect the result and the performance of the algorithm directly.
2. To determine the evaluation function of optimization problems
   In the solving process, the adaptive value is used to evaluate the quality of the solution. When solving problems, therefore, according to the specific characteristics of the problem, the appropriate objective functions must be chosen to calculate the fitness. The fitness is the only parameter to reflect and guide the ongoing optimization process.

3. To choose the control parameters

    The PSO parameters usually include the size of the particle group (the number of particles), the maximum number of iteration algorithm implementation, inertia coefficient, and the parameters of the cognitive, social and other auxiliary control parameters, etc. According to different algorithm models, the appropriate control parameters are selected, and the optimal performance of the algorithm is affected directly.

4. Flight model of particles

    In PSO algorithm, the key is how to determine the speed of the particles. As the particles are described by multidimensional vector, the corresponding flight speed of particles can be described as a multidimensional vector. The speeds and directions of particles will be adjusted along each component direction dynamically in the process of flight by means of their own memories and social sharing information.

5. To determine of the termination criterion of the algorithm

    The most common termination criterion of the particle swarm algorithm is setting a maximum flight algebra in advance, or terminating the algorithm when the fitness in successive generations has no obvious improvement during the search process.

6. Programming operation

    Making program, according to the designed algorithm, and obtaining the solution of the specific optimization problem. The validity, accuracy, and reliability of the algorithm can be verified by evaluating the quality of the solution.

In intelligent control of civil engineering structures by using the PSO algorithm, the above design procedure is usually adopted. The numerical example about the intelligent control of civil engineering structures by using the PSO algorithm will be described in Section 8.5 in detail.

2.12 Genetic Algorithm

GA is an optimized search algorithm proposed based on natural selection and genetic mechanism. Compared to other intelligent control algorithms, such as the Fuzzy Logic and the Neural Network, the GA with self-adapting step size can search the best solution directly and more likely to the global optimum result. The GA provides an effective way to solve complex optimization problem in the field of structural vibration control. Moreover, the research on the combined application of the GA, the Fuzzy Logic and the Neural Network has drawn the attention of many researchers due to their complementary features.

2.12.1 Basic Principle

The GA is a search process based on natural selection and genetics, and usually consists of three operations: Selection, Genetic Operation, and Replacement, as shown in Fig. 2.4 [64].

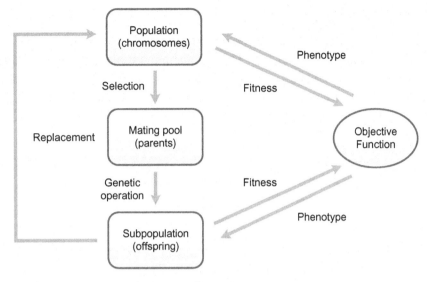

Figure 2.4
Schematic diagram of the GA cycle.

A group of chromosomes constitutes the population of the GA cycle, and can be selected as candidates for the solutions of problems. Firstly, a population is generated randomly, and the objective function in a decoded form is calculated to analyze the fitness values of all chromosomes. Then a set of initial chromosomes is selected as the parents to generate offspring, according to the specified genetic operations, and a same fashion to the initial population is used to evaluate the fitness values of all the offspring. Finally, all the offspring will replace current chromosomes according to a certain replacement strategy.

The GA cycle will be performed repeatedly and terminated when a desired criterion (e.g., a predefined number of generations is produced) is reached. If all goes well throughout the simulated evolution process, the best chromosome in the final population can become a highly evolved solution.

2.12.2 Procedure of GA

Various techniques are employed in the GA process for encoding, fitness evaluation, parent selection, genetic operation, and replacement, which are introduced below.

2.12.2.1 Encoding scheme

The encoding scheme is a key issue in any GA because of its severe limitation on the window of information observed from the system [65]. In general, a chromosome representation, in which problem specific information is stored, is desired to enhance the performance of the algorithm. The GA evolves a multiset of chromosomes. The chromosome is usually expressed in a string of variables in binary [66], real number [67,68] or other forms, and its range is usually determined by specifying the problem.

2.12.2.2 Fitness techniques

The mechanism for evaluating the status of each chromosome can be mainly provided by the objective function, whose values vary from problem to problem. Therefore a fitness function is needed to map the objective value to a fitness value so that the uniformity over various problem domains can be maintained [64]. A number of methods can be used to perform this mapping, and two commonly used techniques are given as follows:

1. Windowing

    Each chromosome can be assigned with a fitness value f_i, which is proportional to the "cost difference" between chromosome i and the worst chromosome.
    In mathematics, it is expressed as

    $$f_i = c \pm (V_i - V_w) \quad (2.151)$$

    where V_i is the objective value of chromosome i; V_w is the objective value of the worst chromosome in the population; and c is a constant. The positive and negative signs in Eq. (2.151) are appropriate for the maximization and minimization problems, respectively.

2. Linear normalization

If the function is to be maximized or minimized, the chromosomes will be ranked in descending or ascending order according to their objective values. Given a raw fitness f_{best} to the best chromosome, the fitness of ith chromosome can be derived from the following expression

$$f_i = f_{\text{best}} - (i - 1) \cdot d \quad (2.152)$$

where d is the decrement rate.

This technique ensures that the average objective value of the population is mapped into the average fitness.

2.12.2.3 Parent selection

Parent selection, which emulates the survival-of-the-fitted mechanism in nature, means that a fitter chromosome can receive a higher number of offspring and thus has a higher chance of surviving in the subsequent generation [64]. Although in many ways, such as ranking, tournament, and proportionate schemes can be used to achieve effective selection [69], the key assumption is to give preference to fitter individuals.

For example, in the proportionate scheme, the growth rate tsr of chromosome x with a fitness value $f(x, t)$ can be defined as:

$$tsr(x, t) = \frac{f(x, t)}{F(t)} \quad (2.153)$$

where $F(t)$ is the average fitness of the population.

2.12.2.4 Genetic operation

Crossover is a recombination operator that combines the subparts of two parent chromosomes to produce offspring that contain some parts of both parents' genetic material [64]. The crossover operator is deemed to be the determining factor that distinguishes the GA from all other optimization algorithms.

A number of variations on crossover operations are proposed, such as one-point crossover, multipoint crossover, and mutation. The diagram illustration of genetic operation is shown in Fig. 2.5.

Mutation is an operator that introduces variations into the chromosome. The operation occurs occasionally, but randomly alters the value of a string position.

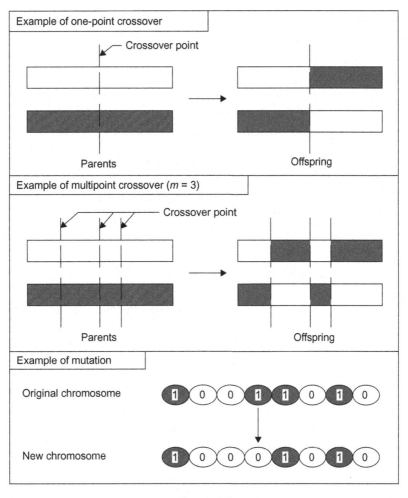

Figure 2.5
Example diagram of genetic operation.

2.12.2.5 Replacement strategy

Two representative strategies [64] can be used for old generation replacement after offspring generation:

1. Generational replacement
   Each population of size n generates an equal number of new chromosomes to replace the entire old population.
2. Steady-state reproduction
   Only a few chromosomes are replaced once in the population. In general, the new chromosomes inserted into the population will replace the worst chromosomes to produce succeeding generation.

Generally, the typical implementation procedure of the GA can be summarized as:

1. Randomly generate an initial population $X(0) = (x_1, x_2, \ldots, x_N)$;
2. Compute the fitness $f(x_i)$ of each chromosome x_i in the current population $X(t)$;
3. Create new chromosomes $x_r(t)$ by mating current chromosomes and applying mutation and recombination as the parent chromosomes mate;
4. Delete numbers of the population to make room for the new chromosomes;
5. Compute the fitness of $x_r(t)$, and insert them into population;
6. $t = t + 1$, if not (end-test) go to step 3, or else stop and return the best chromosome.

Although the global optimum solution can be found easily using the GA, some derivative algorithm, such as the hybrid GA and the cooperative co-evolutionary GA, have been proposed because of its poor local search optimization ability and premature convergence and random walk, and are gradually replacing the traditional GA.

2.12.3 GA Control Realization

In general, the GA is used in combination with other intelligent algorithms in the field of structural vibration control. For example, the GA is widely used as an effective additional feedback algorithm in neural network models and is generally used to find optimal parameters for the fuzzy logic control [70].

The membership functions of input and output variables and the fuzzy control laws can be optimized using GA with the decimal encoding system of the gene. The GA based on the hierarchical structure can optimize weights (including nodes threshold) of neural network with a high learning efficiency. The global search of GA can be used to optimize the fuzzy neural network's parameters off-line. The capability of parallel search with the GA can be used to dynamically optimize the structural parameters of the Back Propagation (BP) network.

The GA optimization approaches for determination of the near-optimal layout of control devices and sensors have also been investigated to improve the active control efficiency of the civil structural system [70]. In practice, the GA is used to find optimal control forces for any time step under earthquake excitation, and dynamic fuzzy wavelet neuro emulator is created to predict structural displacement responses from immediate past structural response and actuator dynamics [71].

CHAPTER 3

Active Intelligent Control

Active intelligent control was started in the 1950s, but the systemic research and practical application in civil engineering structures was started around 1990. The most common active control devices are active mass damper (AMD) system and active tendon control system. AMD system and active tendon control system will be introduced in detail based on the generally accepted principles and classifications.

3.1 Principles and Classification

3.1.1 Buildup of Systems

As described in Chapter 1, Introduction, active intelligent control system consists of sensor system, control system, actuator system, and other components. Sensor system measures the external disturbances or structural response information, and transmits them to the control system. The control system calculates the required active control forces based on a given control algorithm, and outputs control signals to the power drive system (converted by the control force through the control loop), then provides desired control forces on the structure to reduce the responses of the structure with the help of an external energy source. Buildup of systems includes buildup of the main frame and each system, and the main frame is generally built as shown in Fig. 1.9.

In the buildup of active intelligent control system, optimal control forces applied to structures are required to be determined in accordance with different working modes. There are two kinds of working modes, i.e., feedforward control (installing sensors on the foundation) and feedback control (installing sensors on the inner structure). The feedforward control mode is relatively simple, and the control forces can be adjusted real time in accordance with the excitation information, while the responses of structures are not reflected in the control process. Open-loop control system is the common control form for the feedforward control mode, and the schematic diagram is shown in Fig. 3.1(A). The feedback control mode is adjusting the control forces real time in accordance with responses of structures. Closed-loop control system is the common control form of the feedback control mode, and the schematic diagram is shown in Fig. 3.1(B). In addition, the two working modes can also work simultaneously by adjusting the control force, according

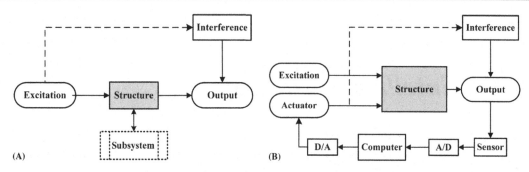

Figure 3.1
Controller working mode. (A) Open-loop control system and (B) closed-loop control system.

to outside interference and comprehensive information, in order to achieve the goal of reducing structural dynamic responses.

3.1.2 Basic Principles

As described in Section 2.1, the equation of motion of the active intelligent control system can be written as Eq. (2.1), and the state-space equation can be expressed as Eq. (2.5), the control force $\{f_d(t)\}$ can be expressed as

$$\{f_d(t)\} = -[G_b]\{Z(t)\} - \{G_f\}\{P(t)\} \tag{3.1}$$

where $[G_b]$ and $[G_f]$ are the feedback gain matrix of system state and a feedforward gain matrix of outside interference, respectively. Substituting Eq. (3.1) into Eq. (2.5), it can be obtained as

$$\{\dot{Z}(t)\} = ([A] - [B_0][G_b])\{Z(t)\} + ([D_0] - [B_0][G_f])\{P(t)\} \tag{3.2}$$

Comparing Eq. (2.5) and Eq. (3.2), it can be seen that the values of structural dynamic property and exciting force change from $[A]$ and $[D_0]$ to $\{[A] - [B_0][G_b]\}$ and $([D_0] - [B_0][G_f])$ because of the action of control force. From the above result, it can be seen that the ideal controlled state of structure will be realized if the increment can be continuously and actively changed to keep it in the optimal state at all time. The modern cybernetics methods described in Chapter 2, Intelligent Control Strategies, can be adopted to realize the goal of ideal active control.

3.1.3 Classification

The active control system shown in Fig. 1.9 can be divided into three kinds of control modes according to the working mode of the controller.

1. Open-loop control

    The control system adjusts the active control force, according to the external excitation information.

2. Closed-loop control

    The control system adjusts the active control force, according to the structural response information.

3. Open/closed-loop control

    The control system adjusts the active control force, according to the integrated information of the external excitation and the structural response.

The commonly used active, intelligent control devices include the following:

1. Active mass damper

    The AMD is the most effective device to control the dynamic responses of high-rise buildings and slender civil structures, which mainly consists of tuned mass damper (TMD) and actuator.

2. Active mass driver

    The active mass driver is similar to the AMD, but it requires no tuning of the primary structure, and the stroke demand on the actuator is less than that of the AMD.

3. Active tendon system

    The active tendon system (ATS) generally consists of prestressed tendons connected to structures and actuators, and the control force is always provided by electrohydraulic servomechanisms.

4. Aerodynamic appendage

    Aerodynamic appendage is a control device for the tall buildings under wind load, which applies the wind energy to eliminate the external energy.

5. Pulse generator

    Pulse generator always means the gas pulse generator, the control force is provided by the release of air jets based on pulse control algorithms.

3.2 Active Mass Control System

3.2.1 Basic Principles

AMD is one of the most effective device to control the dynamic responses of high-rise buildings and slender civil engineering structures, which is derived from the passive TMD. The use of actuators and active control algorithms in AMD control system overcome some drawbacks of TMD: the sensitivity to tuning error, the time interval of full operation, and incapacity in high mode [72].

The AMD control system commonly consists of sensor, controller, and AMD device. The basic control frames of AMD control system are shown in Fig. 1.9. When the AMD control system is working, the sensors collect information of outside interference or (and) response of the structure, and give a feedback to the controller. Then the active control force is calculated by the controller in accordance with a kind of active control algorithm. Finally, the actuator drives the inertial mass and applies the control force on structures. So, the control of structural vibration is achieved effectively.

Generally, the AMD system includes inertial mass, stiffness element, damping element, and actuator connected to the top of the structure as shown in Fig. 3.2. Figs. 3.2 (A)–(C) show the standard AMD system, AMD system without stiffness and damping elements, and AMD system combined with TMD, respectively [27].

For the standard AMD system, the system is firstly designed as a TMD system without active control actuator. That is to say, under the premise of a given inertial mass, the system should meet the relationships of optimal frequency ratio and optimal damping ratio of a TMD system. Then, based on the designed TMD system, the active control force of the active control actuator is calculated according to the active control algorithm. This design method of the AMD system has two advantages:

1. The AMD system has the reliability of fail-safe: the TMD system will display the passive control effect when the active control actuator fails or stops working.
2. In larger environment disturbance, the active actuator is working while actually the AMD is a system with spring and damper, which has the better control effect and higher control robustness compared to the TMD system. Moreover, when the

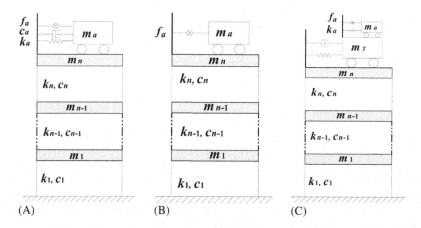

Figure 3.2
The forms and calculation modes of AMD structural control system. (A) The standard AMD system, (B) the AMD system without stiffness and damping elements, and (C) the AMD system combined with TMD.

environmental disturbance is small, the working state changes from AMD to TMD, so the active actuator stop working to save energy and extend the life of the active control actuator.

3.2.2 Construction and Design

As obviously seen, the construction of a standard AMD system should include four parts, i.e., the inertial mass, the stiffness element, the damping element, and the actuator. Here, the design of a standard AMD system installed on a single freedom structure will be introduced as an example. Fig. 3.3 shows a standard AMD system, the equations of motion of the controlled single-freedom structure can be expressed as

$$m\ddot{x}(t) + c\dot{x}(t) + kx(t) = -m\ddot{x}_g(t) + p(t) + f_d(t) \tag{3.3}$$

$$m_a(\ddot{x}_a(t) + \ddot{x}_g(t)) + c_a(\dot{x}_a(t) - \dot{x}(t)) + k_a(x_a(t) - x(t)) = f_a(t) \tag{3.4}$$

where m, c, and k are the mass, damping coefficient, and stiffness of the structure, respectively. $p(t)$ is the disturbance of the system. $x(t)$, $\dot{x}(t)$, and $\ddot{x}(t)$ separately are the displacement, velocity, and acceleration of structure relative to ground. $\ddot{x}_g(t)$ is the acceleration of the ground. m_a, c_a, and k_a separately are the mass, damping coefficient, and stiffness of the AMD system. $x_a(t)$, $\dot{x}_a(t)$, and $\ddot{x}_a(t)$ separately are the displacement, velocity, and acceleration responses of the AMD system relative to ground. $f_a(t)$ are the driving force applied to the inertial mass by the actuator. $f_d(t)$ is the active control force provided by the AMD system to the controlled structure, which can be written as:

$$f_d(t) = c_a(\dot{x}_a(t) - \dot{x}(t)) + k_a(x_a(t) - x(t)) - f_a(t) = -m_a(\ddot{x}_a(t) + \ddot{x}_g(t)) \tag{3.5}$$

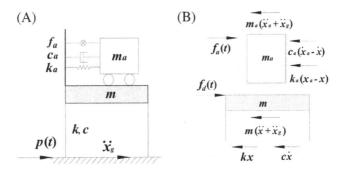

Figure 3.3
Single-freedom controlled structure with AMD system and force analysis. (A) The AMD structural control system and (B) force analysis of the controlled system.

From Eq. (3.3) to Eq. (3.4), the equations of motion of the AMD control system can be rewritten as:

$$\begin{bmatrix} m & 0 \\ 0 & m_a \end{bmatrix} \begin{Bmatrix} \ddot{x}(t) \\ \ddot{x}_a(t) \end{Bmatrix} + \begin{bmatrix} c+c_a & -c_a \\ -c_a & c_a \end{bmatrix} \begin{Bmatrix} \dot{x}(t) \\ \dot{x}_a(t) \end{Bmatrix} + \begin{bmatrix} k+k_a & -k_a \\ -k_a & k_a \end{bmatrix} \begin{Bmatrix} x(t) \\ x_a(t) \end{Bmatrix}$$
$$= -\begin{Bmatrix} m \\ m_a \end{Bmatrix} \ddot{x}_g(t) + \begin{Bmatrix} 1 \\ 0 \end{Bmatrix} p(t) + \begin{Bmatrix} -1 \\ 1 \end{Bmatrix} f_a(t) \tag{3.6}$$

It can be seen from the right part of Eq. (3.5) that the active control force applied to the structure of the standard AMD system includes the driving force, the damping force, and the elastic force; the stiffness and damping elements of AMD can limit the displacement of inertial mass to some level, and output force is prorated to each device in the vibration process. In addition, the active control force is equal to the inertial force of inertial mass. Therefore the vibration control effect almost depends on the weight of the inertial mass. Generally, it will get the better vibration control effect for taking inertial mass with larger weight.

It can be seen from Eqs. (3.5) and (3.6) that there are four important design parameters for the AMD control system: the driving force, the inertial mass, the damping coefficient, and the stiffness.

As mentioned in Section 3.2.1, there are two steps for designing the AMD control system. In the first step, assuming the active control actuator does not work and the control system can be designed as a TMD system. In the whole system, the structure mass m, stiffness k, and damping coefficient c can be calculated, and the natural frequency of the TMD system is $\omega_a = \sqrt{\frac{k_a}{m_a}}$, the damping ratio is $\zeta_a = \frac{c_a}{2 m_a \omega_a}$. Setting $\mu = \frac{m_a}{m}$ is the mass ratio of the AMD system and the structure, then these parameters of the control system m_a, k_a, and c_a can be calculated through Eq. (3.7) [73]:

$$\beta_a = \frac{\omega_a}{\omega_0} = \frac{\sqrt{1-\frac{\mu}{2}}}{1+\mu} \quad \zeta_a = \sqrt{\frac{\mu\left(1-\frac{\mu}{4}\right)}{4(1+\mu)\left(1-\frac{\mu}{2}\right)}} \tag{3.7}$$

Secondly, the driving force $f_a(t)$ should be designed according to a particular control algorithm. The two ends of Eq. (3.5) are integrated by $d(x_a - x)$, then

$$\int f_a(t) d(x_a - x) = \int -m_a(\ddot{x}_a + \ddot{x}_g) d(x_a - x)$$
$$= \int [c_a(\dot{x}_a - \dot{x}) + k_a(x_a - x) - f_a(t)] d(x_a - x) \tag{3.8}$$
$$= \int c_a(\dot{x}_a - \dot{x}) d(x_a - x) + 0.5 k_a(x_a - x)^2 - \int f_a(t) d(x_a - x)$$

It can be seen from Eq. (3.8) that the work done by the control force applied to the structure is related to the motion of the inertial mass. That is to say, a part of vibration energy of the structure will be converted to the kinetic energy of AMD system. Eventually, the control force applied to the structure will balance the vibration energy to achieve the purpose of structural vibration control. In addition, the work done by the control force is equal to the difference between the external energy and the energy of damping, elastic deformation of AMD system. That is to say, the damping element and the stiffness element of AMD system will consume and absorb a part of vibration energy of AMD system or external energy. Therefore the control effect can be improved by increasing the external energy import.

If there are no stiffness element and damping element in the AMD system, the equations of motion of the AMD control system can be expressed as

$$m_a(\ddot{x}_a(t) + \ddot{x}_g(t)) = f_a(t) \tag{3.9}$$

In addition, the control force applied on the structure is

$$f_d(t) = -m_a(\ddot{x}_a(t) + \ddot{x}_g(t)) = -f_a(t) \tag{3.10}$$

Eq. (3.10) shows that when the AMD system has only inertial mass without stiffness element and damping element, the driving force of the AMD actuator equals to the control force, also exactly equals to the inertia force of the inertial mass.

3.2.3 Mathematical Models and Structural Analysis

In the AMD control system, the key point is how to calculate the control force $f_d(t)$ and make it close to the driving force $f_a(t)$, which involves the control strategy and control device design. The AMD control system is installed on the top of a structure with n degrees of freedom as shown in Fig. 3.4. Here, the $\{P(t)\}$ means the earthquake acceleration $\{\ddot{x}_g(t)\}$ and the wind load $\{F_w(t)\}$, the equation of motion of the AMD structural control system can be expressed as

$$[M]\{\ddot{x}(t)\} + [C]\{\dot{x}(t)\} + [K]\{x(t)\} = -[M][\Gamma]\{\ddot{x}_g(t)\} + [D_w]\{F_w(t)\} - [B]\{f_d(t)\} \tag{3.11}$$

where $[M]$, $[C]$, and $[K]$ separately are the mass matrix, damping matrix, and stiffness matrix of AMD control system. $\{x(t)\}$, $\{\dot{x}(t)\}$, $\{\ddot{x}(t)\}$ separately are the displacement vector, velocity vector, and acceleration vector of the AMD control system. $[B]$ is the position matrix of AMD control force. $\{f_d(t)\}$ is the AMD control force vector, which is equal to $\{f_a(t)\}$ in this model. $[D_w]$ is the position matrix of the wind-load vector.

Assuming $\{Z(t)\} = \{\{x(t)\}, \{\dot{x}(t)\}\}^T$, then the state-space equation of the motion of the AMD control system can be expressed as

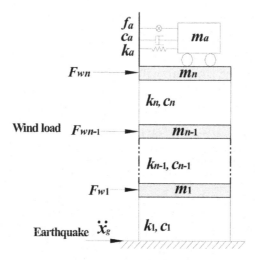

Figure 3.4
The AMD structural control system.

$$\begin{cases} \{\dot{Z}(t)\} = [A]\{Z(t)\} + [B_0]\{f_d(t)\} + [D_w]\{F_w(t)\} + [D_g]\{\ddot{x}_g(t)\} \\ \{Y\} = [E_0]\{Z(t)\} \end{cases} \quad (3.12)$$

where $A = \begin{bmatrix} [0] & [I] \\ -[M]^{-1}[K] & -[M]^{-1}[C] \end{bmatrix}$, $[B_0] = \begin{bmatrix} [0] \\ [M]^{-1}[B] \end{bmatrix}$, $[D_w] = \begin{bmatrix} [0] \\ [M]^{-1}[I] \end{bmatrix}$,

$[D_g] = \begin{bmatrix} [0] \\ -[\Gamma] \end{bmatrix}$. $[E_0]$ is the measurement matrix, $\{Y\}$ is the measurement vector.

In Eqs. (3.11) and (3.12), $\{f_d(t)\}$ actually is the driving force of AMD system, that is to say, structural vibration control can be realized by actively changing the driving force in the design process. Here, taking $\{f_d(t)\}$ to express the driving force $\{f_a(t)\}$. The driving force $\{f_d(t)\}$ in Eq. (3.12) can be calculated by any kind of active control algorithm described in Chapter 2, Intelligent Control Strategies. Here, the well-known linear quadratic regulator (LQR) is chosen as the active control algorithm, and the corresponding performance index including response of structure and auxiliary mass is given by

$$J = \int_0^\infty (\{Z(t)\}^T[Q]\{Z(t)\} + \{f_d(t)\}^T[R]\{f_d(t)\}) \, dt \quad (3.13)$$

where $[Q]$ and $[R]$ separately are the positive semi-definite and positive-definite weight matrices, which are the control parameters in the AMD system design. The optimal driving force, minimizing the performance index based on LQR algorithm is expressed as:

$$\{f_d(t)\} = -[G]\{Z(t)\} \quad (3.14)$$

where $[G] = [R]^{-1}[B_0]^T[\tilde{P}]$ is the gain matrix, $[\tilde{P}]$ is the positive-definite solution of the following matrix Riccati equation:

$$[\tilde{P}][A] + [A]^T[\tilde{P}] - [\tilde{P}][B_0][R]^{-1}[B_0]^T[\tilde{P}] + [Q] = 0 \qquad (3.15)$$

So, the state-space equation of the AMD control system is transformed to

$$\{\dot{Z}(t)\} = ([A] - [B_0][G])\{Z(t)\} + [D_w]\{F_w(t)\} + [D_g]\{\ddot{x}_g(t)\} \qquad (3.16)$$

Eq. (3.16) can be solved by any kind of the numerical methods of solving linear differential equations, such as *Wilson* $-\theta$ method, *Newmark* $-\beta$ method. However, there is a more commonly used method, such as the differential equation solver of function *lsim* and function *ode* in MATLAB, to solve the state vector of AMD control system.

Eq. (3.14) can be rewritten as:

$$\{f_d(t)\} = -[K_G]\{x(t)\} - [C_G]\{\dot{x}(t)\} \qquad (3.17)$$

where $[K_G]$ and $[G_G]$ are the sub-matrices of the gain matrix $[G]$. Eq. (3.17) shows that the AMD control system applies elastic force and damping force on the structure, that is to say, the AMD control system actually decreases the structural dynamic response by changing the stiffness and damping of the structure. Moreover, a set of AMD device is typically installed on the top of the structure, it can be observed that the AMD control system only changes the nth row elements of the stiffness and the damping matrices by taking Eq. (3.17) into Eq. (3.11).

The design of the AMD control system includes two aspects: one is the design of AMD system parameters, such as inertial mass m_a, damping coefficient c_a, stiffness k_a, and so on, which have been discussed in Section 3.2.2. The other one is AMD control parameter design, such as the maximum driving force and the maximum stroke of inertial mass. These two kinds of parameters are usually coupled, and the latter depends on some particular parameters of the control algorithm (e.g., the weight matrix $[Q]$ and $[R]$ of LQR algorithm).

The driving force of the AMD control system can be calculated by LQR algorithm, as seen in Eqs. (3.13) and (3.14), and the weight matrix $[Q]$ and $[R]$ greatly influence the driving force, the structural response, and the AMD inertial mass strike. Generally, the greater the $[Q]$ is, and the smaller the $[R]$ is, the smaller the structural responses and the greater of driving forces. Moreover, in order to make sure the AMD control system can effectively work, the weight matrix $[Q]$ and $[R]$ should be designed for different disturbance levels, which avoids to be damaged due to large stroke of inertial mass and large driving force under the strong disturbance.

The purpose of designing the control device is to provide enough driving force to be close to the desired control force. However, the different control devices with different working principle can provide different control effects to the AMD control system. The typical forms of AMD control device are pendulum, rubber pad, rail-type, and so on. Similarly, for these devices of rubber pad and rail-type, equations of motion and control forces which can be obtained from Eq. (3.4) to Eq. (3.5) or from Eq. (3.9) to Eq. (3.10) according to the different working principles.

3.2.4 Experiment and Engineering Example

AMD was proposed to control the structural vibration in 1980 [29]. In the past three decades, more and more experiments and applications of the AMD control system were conducted.

In 1987, Aizawa conducted a shake-table experiment of a four-story model fame installed an AMD, and achieved satisfactory control effect [74]. An active control experiment of AMD system that placed on top of a three-story model steel frame was also conducted by Kobori in 1987 [75]. In the work of Soong in 1988, an AMD was placed on top of a 1:4 scale six-story model steel fame, and the active control experimental results showed the effectiveness of the AMD control system [76]. On the other hand, the first engineering application of the AMD control system in the world was conducted in Japan in 1989 [77], the controlled structure is 11-story office building and the AMD is designed to control the lateral and torsional responses of structure acted by wind load and earthquake. Subsequently, AMD control systems are applied into a lot of high-rise buildings and slender civil structures, especially in Japan.

As a typical case of the engineering project, the wind vibration problem of Nanjing TV tower has been solved by using the AMD control system for the first time in China [78−79]. Nanjing TV tower, height 310.1 m, is located in the western suburbs of Nanjing city, the main structure is made up of three prestressed concrete hollow legs with rectangular section as shown in Fig. 3.5. There are two sightseeing platforms located in the tower at the heights of 200 m and 240 m, and the masses of the two platforms are 10000 t and 1000 t, respectively. The analysis results showed that the periods of the tower for the 1st, 2nd, and 3rd mode shape separately are 6.20 s, 2.10 s, and 1.14 s. The bearing capacity and the horizontal displacement of the tower can meet the requirements under the fundamental wind pressure of 0.5 kN/m^2 or earthquake intensity of 8 degree. However, the acceleration calculated is 0.2 m/s^2 which exceeds the human body comfort level limit 0.15 m/s^2. Therefore wind vibration control is needed to conduct to meet the requirement of comfort level. Considering the stiffness above the big platform is far less than the stiffness of the main structure, applying control force on the small platform is a good choice to get a better control effect.

The AMD control system layout of Nanjing TV tower is shown in Fig. 3.6. An electro-hydraulic servo actuator is arranged in the annular site each angle of 120 degree, with the

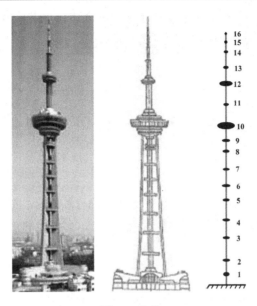

Figure 3.5
Nanjing TV tower and its simplified analysis model.

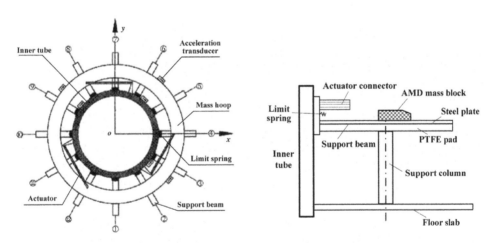

Figure 3.6
The AMD control system arrangement of Nanjing TV tower.

length of 6.5 m, the maximum stroke of 750 mm and the maximum output force of 100 kN. Thus the three driving forces provided by actuators can form two horizontal forces and a torque, so the translation and rotation of TV tower can be controlled successfully. However, the dynamic response of Nanjing TV tower is dominated by the first mode shape, so the AMD system is used to reduce the wind vibration of the first mode shape of the TV tower,

Table 3.1: The analysis results of the small turret

Conditions	Displacement (x/mm)	Acceleration ($\ddot{x}$/mg)	Diving Force (F_a/kN)
Uncontrolled	137	31.8	—
AMD control	89	16	157.2

and the driving force is calculated according to the LQR control algorithm. The final analysis results are shown in Table 3.1.

It can be seen from Table 3.1 that the displacement and acceleration of the small platform are all obviously reduced by using the AMD control system. The controlled result meets the requirement of human body comfort level. So, the AMD control system used in Nanjing TV tower can achieve the desired performance target. Besides, the driving force is not very large, which proves the high efficiency of AMD control system. Recently, Zhou conducted a hybrid mass damper (HMD) control strategy to control the wind-induced vibration of Canton Tower. The HMD consists of two AMD subsystems driven by multiple linear motors with maximum capacity of 18 kN, and some measures are taken to consider the velocity limitation and stroke limitation of AMD. The actual test results show the AMDs can implement the control force synchronously and have a very small time delay which can meet the requirement of vibration control for the Canton Tower [80].

3.3 Active Tendon System

3.3.1 Basic Principles

ATS was proposed by Freyssinet in 1960, which is one of the most widely used active control methods applied in flexible structures, high buildings, and bridges [81]. Active tendon consists of prestressed tendons and actuators, and the prestressed tendons with cross form are always installed between every pair of neighboring floors. The electrohydraulic servomechanism generates the driving force to change the tension of prestressed tendons; as a result, the active control force generated by tendons is applied to the controlled structure. A lot of tests and simulations show the ATS has an excellent control effect on horizontal vibration or torsional vibration of structures.

Actually, the control principle of ATS is similar to AMD, when the structure is subjected to earthquake or wind loads, the structure will deviate from the equilibrium position. Meanwhile, sensors installed on and near the structure measure the information of structural response and external excitation, and transmit signals to the controller. The controller calculates the active control force, according to a particular active control algorithm, and commands the actuator to change the tension of tendons. Finally, some horizontal control forces are applied to the structure, so that the control of structural vibration is achieved effectively.

Besides the excellent control effect, the ATS has some special advantages. Existing tendons can be used or modified as active tendons that few additions or modifications need to be done for an as-built structure, and the additional installation space is unnecessary because of making use of the existing structural members, which is attractive for the efficiency retrofits and strengthening of existing structures [42,81]. On the other hand, active tendon control can accommodate both continuous-time and pulse control algorithms as the ATS can be conducted in the pulsed mode and the continuous-time mode [42,81]. Like other active control mechanisms, ATS has the same drawback that the actuator needs to be supported by a strong external energy, in addition, it is more sensitive to the delay.

3.3.2 Construction and Design

Active tendon consists of prestressed tendons and actuators, the commonly used actuator is electro-hydraulic servo actuator, the prestressed tendon has oblique form and cross form according to the arrangement of the tendon, the widely used ATS in civil engineering is shown in Fig. 3.7.

Fig. 3.7(A) shows the ATS designed by Soong et al. for controlling the structural vibration under seismic loads in 1988, which is also the first generation Benchmark model for structural vibration control [81,82]. Two pairs of diagonal prestressed tendons are installed on the first layer of three-story steel frame, and a connecting frame is used to connect electro-hydraulic servo actuator and the steel frame.

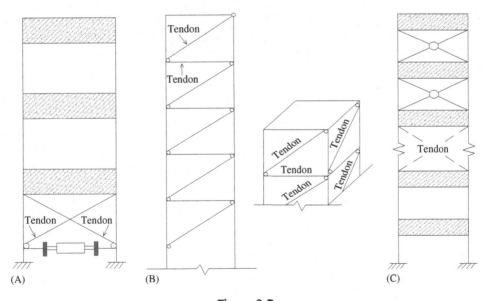

Figure 3.7
The common ATS in civil engineering. (A) Designed by Soong; (B) Designed by Abdel-Rohman; (C) Designed by Yang.

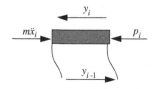

Figure 3.8
The stress analysis of a story unit.

Fig. 3.7(B) shows the ATS designed by Abdel-Rohman for controlling the structural vibration under wind loads in 1983 [30]. The device consists of four prestressed tendons connected to the top corners of the structure, one pass down through pulleys on each side of structure to cover the most flexible region. Four actuators are installed at the certain level to connect the four prestressed tendons, so that the active control can be achieved through tightening or releasing the prestressed tendons.

Fig. 3.7(C) shows the ATS designed by Yang for controlling the structural vibration under wind loads in 1983 [83]. This device combines the features of the above two ATSs, the diagonal prestressed tendons, controllers, and actuators are installed in the space between adjacent floors, the arrangement starts from the top of the structure, and ends in a certain level, ensuring that the most flexible region of structure is controlled.

The determination of the active control force depends on the mathematical model of ATS and the active control algorithm selected. Here, the ATS in Fig. 3.7(C) is chosen as an example to explain the design of active control force, and the stress analysis of a story unit with a pair of prestressed tendons is shown in Fig. 3.8, and the corresponding equation of motion and force-displacement relation are given by

$$m_i \ddot{x}_i + y_{i-1} - y_i - p_i = 0 \tag{3.18}$$

$$y_{i-1} = k_i(x_i - x_{i-1}) + c_i(\dot{x}_i - \dot{x}_{i-1}) + f_i \tag{3.19}$$

where m_i, c_i, and k_i are, respectively, the mass, damping coefficient, and stiffness of ith story unit. x_i, $\dot{x}_i$, and $\ddot{x}_i$ are, respectively, the displacement, velocity, and acceleration of ith story unit relative to the ground. y_{i-1} is the resultant shear force in all the columns of ith story unit, f_i is the active control force provided by prestressed tendons of ith story unit.

Substituting Eq. (3.19) into Eq. (3.18), the equation of motion of ith story unit can be rewritten as

$$m_i \ddot{x}_i + c_i(2\dot{x}_i - \dot{x}_{i-1} - \dot{x}_{i+1}) + k_i(2x_i - x_{i-1} - x_{i+1}) = p_i - (f_i - f_{i+1}) \tag{3.20}$$

Considering the force mechanism of every story unit is similar, all the equations of motion are assembled as the equation of the system, and the active control force in the equation of motion can be deleted for the story unit without prestressed tendons. Introducing the boundary conditions of real structure, the equation of motion of the controlled structure with ATS can be written as Eq. (2.1), and $\{f_d(t)\}$ is the control force vector provided by prestressed tendons. In order to use active control algorithm to conduct the active control, the equation of motion of the controlled structure is rewritten as Eq. (2.5).

Here, the active control force is designed based on pole assignment method described in Section 2.3, and the control algorithm adopts the full-state feedback approach, so the active control force can be expressed as

$$\{f_d(t)\} = -[G]\{Z(t)\} \tag{3.21}$$

where $[G]$ is state feedback gain matrix. Substituting Eq. (3.21) into Eq. (2.5), and the external excitation is ignored, then the state-space equation can be rewritten as

$$\{\dot{Z}(t)\} = ([A] - [B_0][G])\{Z(t)\} \tag{3.22}$$

where the gain matrix $[G]$ is calculated according to the steps described in Section 2.3. $[G]$ is designed as

$$[G] = -[e][\Gamma]^{-1} \tag{3.23}$$

The state of the controlled structure is calculated according to the above equations using function *ode* in MATLAB. However, the control force is just the optimal force applied to the controlled structure in the horizontal direction, not the tension force derived from the prestressed tendons. As the prestressed tendons are arranged in the space between adjacent floors with some certain angles, the force needs to be decomposed according to geometry knowledge. On the other hand, the operation of the actuator that changing the tension force of prestressed tendon is based on the relationship between force and axial deformation of prestressed tendon, so the establishment of the relationship is essential, and the initial prestress of prestressed tendon should also be considered. After the above considerations and operations, the ATS can control the structural vibration in practice.

3.3.3 Experiment and Engineering Example

ATS, as an effective method for structural vibration control, is popular in civil engineering due to easy operation and simple construction, and experiments about ATS were conducted in 1980s. Some control experiments of ATS for some small-scale structural models were performed by Roorda in 1980 [84], the effect of time delay on active control effect was studied, and the problems about the actuator dynamics and placement of sensors were also proposed. From 1986 to 1989, the experiments of ATS were conducted systematically by Soong et al., which consisted of three stages, single-story steel frame, three-story steel frame, and six-story steel frame models, control force is applied by an electrohydraulic servomechanism [85], the experiment results of adopting LQR algorithm and independent modal control were also compared and were both satisfactory. In 2001, the vibration test of cable-stayed bridge model with active piezoelectric tendons was conducted by Bossens and Preumont to demonstrate the control effect of ATS, and a large-scale mockup equipped with hydraulic actuators was tested to demonstrate the practical implementation of ATS [85]. A vibration measurement of a beam with a cable mechanism and the motor was conducted by Nudehi et al. in 2006, the results showed the first two modes of the beam can be effectively controlled [86]. A control experiment on a frame structure with cable actuators was tested by Issa et al. to verify the effectiveness of the control scheme in 2010 [87]. However, ATS requires much more energy and a plurality of actuators that make it difficult to be implemented in an engineering project.

Soong et al. and Takenaka Company in Japan installed an ATS in a full-scale dedicated test structure in Tokyo in 1990, in order to control the structural dynamic response under artificial excitation and actual ground motion [88]. The test structure is a symmetric two-bay, six-story building with the mass of 600 t as shown in Fig. 3.9.

In detail, the structure is a steel frame which is composed mainly of rectangular tube columns, W-shaped beams, and reinforced concrete stab. The fundamental periods in strong direction and weak direction are 1.1 s and 1.5 s, which is the typical case of high-rise

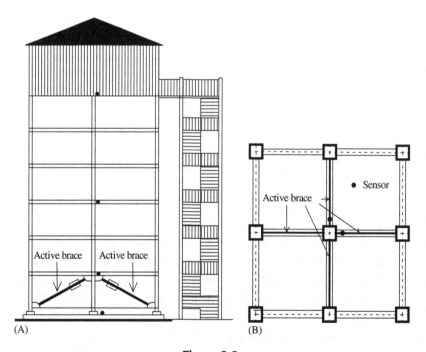

Figure 3.9
The full-scale dedicated test structure in Tokyo. (A) Side view; (B) Top view.

building. The active control system consists of digital controller, sensors, hydraulic servo actuators, and braces, and the diagonal braces are installed in the first story of the building, which can expand and contract to control the structural response. The measured stiffness of the braces in *x*-direction and *y*-direction are 98.4 KN/mm and 73.8 KN/mm.

The active control algorithm adopted the simplified three-velocity feedback control based on linear optimal control theory, which requires smaller control force, but more power is required. The peak responses of the sixth floor are generally reduced in the *x*-direction, but less affected in the *y*-direction, and the root mean square responses are reduced in all cases, which indicates an overall reduction of vibration amplitudes throughout the motion.

3.4 Other Active Control System

Besides the above two main active control systems, some different forms of active control systems are applied in civil engineering, such as an active support system, pulse generator system, and active aerodynamic wind deflector system. The active support system is similar to the ATS, so the related analysis and research are not introduced in the following. The pulse generator system and active aerodynamic wind deflector system will be introduced to display the different working forms of active control systems.

3.4.1 Form and Principles

The pulse generator system is designed to reduce the gradual rhythmic buildup of structural response caused by resonance case [89]. Generally, pulse generator releases air jets to generating short-interval and high-energy pulses, and the magnitude and direction of control pulse depend on the particular control algorithm. Compared to other control systems, it has the advantage of a low-energy input and better real-time operation, and is more effective for critical state control. The pulse generator system usually consists of sensors, controllers, actuators, and pneumatic power supply, Fig. 3.10 shows a pulse generator system designed by Miller in 1988 [89]. The actuator consists of two commercially available solenoids, and the compressed air flow through the nozzle to generate the control pulse. When the structure is subjected to external excitation, the sensors collect the information of structural responses. Once the thresholds at some specified positions are exceeded, the generator will be triggered to generate control pulses. The amplitude of the pulse depends on control algorithm which minimizes the defined cost function; in addition, the direction of control pulse can also be adjusted to achieve a more effective control effect.

Active aerodynamic wind deflector system suppresses wind-induced vibration of the structure by changing the area of windshield to adjust the suffered wind pressure, which meets the requirements of structural safety and human comfort. Obviously, the aerodynamic control needs less energy supply since the active control force is derived from wind power,

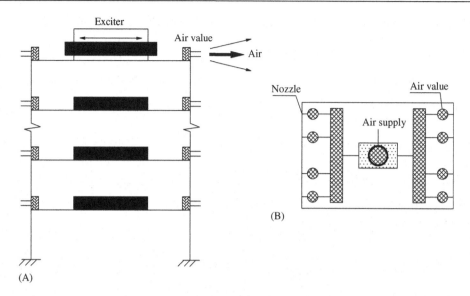

Figure 3.10
Pulse generator system. (A) Side view; (B) Top view.

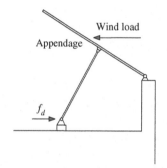

Figure 3.11
Simple aerodynamic appendage.

so the distinct disadvantage is that the control system only controls the vibration induced by wind. Active aerodynamic wind deflector system consists of sensors, controllers, actuators, and appendages (or windshield). Klein designed a simple aerodynamic appendage that reduces wind-induced vibration of tall building, as shown in Fig. 3.11 [90]. Fig. 3.12 shows an aerodynamic flap system designed by Gupta and Soong in 2000 [91]. The working principle of active aerodynamic wind deflector system is similar to the other active control system, but the generation of the control force is special. The actuator adjusts the area and the direction of appendages (or windshield) to exploit the wind power, the suffered wind pressure on the appendages (or windshield) as the control force is applied to the controlled structures.

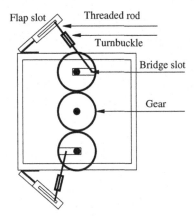

Figure 3.12
Aerodynamic flap system.

3.4.2 Analysis and Tests

For the structure controlled by the pulse generator system, the rhythm buildup of structural responses can be destroyed by the control pulse with suitable magnitude and direction. So, the magnitude and direction of control pulse are the main parameters of the control system. Assuming the structure is subjected to earthquake excitation, the generators are distributed through the controlled structure that the direction of control pulse is determined. When some specified thresholds are exceeded, the pulse generator will be activated, and the optimal control pulse is determined that keeps the cost function minimal over a relatively short time segment T_{opt}. The structural response can be assumed that the stochastic component is superimposed on deterministic component, and the purpose of the control is to minimize the deterministic component, the control pulse vector can be expressed as

$$\{f_{d1}(t), f_{d2}(t), \ldots, f_{dn}(t)\}^T = \{f_d(t)\} = \{a\}f_0(t) \tag{3.24}$$

where $\{a\}$ is the amplitude vector of control pulse, $f_0(t)$ is an arbitrary time history. Neglecting the mean value of the earthquake excitation during the period t_0 to $T_0 = (t_0 + T_{opt})$, the structural responses are given by solving the following equation of motion

$$[M]\{\ddot{x}(t)\} + [C]\{\dot{x}(t)\} + [K]\{x(t)\} = \{f_d(t)\} \tag{3.25}$$

Considering the initial conditions and applying the modal approach, the final structural responses can be written as

$$\{x(t)\} = [G_1]\{x(t_0)\} + [G_2]\{\dot{x}(t_0)\} - [G_4]\{e\} + [G_5]\{f_d(t)\} \tag{3.26}$$

where the matrices are related to the time, and the meaning and the elements in the matrices are mentioned in reference [89]. Because the earthquake excitation is regarded as

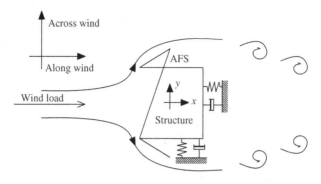

Figure 3.13
Active aerodynamic bidirectional system.

zero-mean random process in T_{opt}, the structural responses do not depend on the earthquake excitation, the variance of displacement responses can be expressed as

$$E[\{x(t)\}] = [G_1]\{x_0\} + [G_2]\{\dot{x}_0\} + [G_5]\{f_d\} \tag{3.27}$$

The final control pulses are determined by selecting the appropriate cost function, the kinetic energy of the controlled structure is selected as the cost function

$$J = E[\{x(t)\}]^T [M] E[\{x(t)\}] dt \tag{3.28}$$

It is worth noting that the above approaches are based on the linear model, the nonlinear characteristic of structure is not considered. The solution based on modal approach is not applicable to solve the nonlinear problem of control pulse, but a very simple and reliable method for nonlinear structure to determine the optimal control pulse is presented by Masri [92].

For the active aerodynamic wind deflector system, the area of the windshield is the main parameter. Here, the active aerodynamic bidirectional system designed by Gupta is used to explain the determination of the area of flaps and angular amplitude of rotation, which is shown in Fig. 3.13.

The system consists of two flaps (or appendages) which rotates about the leading edge of the controlled structure, the exposed area of flaps suffers wind pressure to produce drag forces, so the along-wind motion of structure is controlled. The equation of along-wind motion of structure can be expressed as:

$$\ddot{x} + 2\xi w \dot{x} + w^2 x = \frac{1}{m} \left[\frac{1}{2} \rho A C_D U^2 + \rho A C_D U u(t) + \rho A_P(t) S C_D (U + u(t))^2 \right] \tag{3.29}$$

where $A_P(t)$ is the exposed area of flaps, which is a function of time, S is the control switching function which guides the flap operation, and the meanings of the above symbols are explained in the reference [91]. On the other hand, the system also works as the vertex generator to produce the vortices with the optimal frequency and phase. This feature can be exploited to reduce the potentially hazardous vibration caused by the resonance case, so the across-wind motion of the structure is also controlled. The equation of across-wind motion of structure can be expressed as:

$$\ddot{y} + 2\xi w \dot{y} + w^2 y = \frac{\rho U^2 D}{2m} \left[Y_1(R) \frac{\dot{y}}{U} + Y_2(R) \frac{y^2 \dot{y}}{D^2 U} + B(R) v(t) \frac{y}{D^2} S \right] \quad (3.30)$$

where $v(t)$ is the actuator feedback and the influence of the system on the across-wind motion of the structure, the meanings of the other symbols are also explained in Ref. [91]. In addition, the switch function introduces bidirectional coupling in the two vibratory directions through the control term. The control approach in the across-wind motion is derived from the dissipative effects of parameters, and the control algorithm belongs to the closed-loop feedback algorithm, the above $v(t)$ is the feedback of wake, and the feedback proposed by Fujino et al. [93] can be adopted. The switch function proposed by Klein et al. [90] can be adopted to control the along-wind motion of the structure, and the direction of the sway of controlled structure determines the working condition of the switch function. For bidirectional control of structure, a modified switching function is proposed to achieve a satisfactory control effect [91].

Traina et al. conducted a pulse control on the six-story frame using pneumatic, hydraulic, and electromagnetic actuators, and the experimental results demonstrated the feasibility, reliability, and robustness of the pulse control [94]. Soong and Skinner conducted a wind tunnel experiment on a 1:400 scale elastic model installed aerodynamic appendages, and the result showed a significant reduction in the structural responses [81]. Gupta also conducted a wind tunnel experiment on a rigid model and a flexible model installed active aerodynamic control device, the results showed that the across-wind motion can be controlled effectively, and the assumptions in the mathematical model were also verified [91]. Kobayashi conducted a wind tunnel experiment on the sectional model of a bridge with the control wings, these wings provide aerodynamic forces with adequate phase and amplitude, the test results showed the flutter speed was increased by a factor of two [95]. Nissen adopted an active aerodynamic appendage system to control the wind-induced oscillation in long suspension bridges, the analysis showed that the critical wind speed for flutter instability and the divergence was increased by the active control [96].

CHAPTER 4

Semiactive Intelligent Control

Active control systems rely entirely on external power to operate the actuators and supply the control forces, and can adapt to a wide range of operating conditions and structures. However, in many applications, such active control systems for civil engineering structures require large power source, which makes them costly and vulnerable to power failure, this is why the civil engineering structures are reluctant to use active control systems for vibration mitigation. Another critical issue with active control is the robustness with respect to actuator's failure, and this problem becomes especially serious when centralized controllers are used. On the contrary, semiactive control devices, of which the stiffness and the damping can be adjusted real-time, require less energy than active control devices. The energy can also be stored locally in a battery, thus the semiactive device is independent of any external power supply. Semiactive control systems were proposed as early as the 1920s when patents were issued for shock absorbers, which utilized an elastically supported mass to activate the hydraulic valve or utilized a solenoid valve for directing fluid flow [97]. Over the past years, semiactive control has found its way in many vibration control applications, particularly in structural engineering [98].

This chapter begins with an introduction of semiactive intelligent control. Next, some semiactive control devices, including magnetorheological (MR) damper, electrorheological (ER) damper, piezoelectricity friction damper, semiactive varied stiffness damper, semiactive varied damping damper and magnetorheological elastomer (MRE) device are discussed.

4.1 Principles and Classification

4.1.1 Basic Principles

As described in Chapter 1, Introduction, a semiactive control system may be defined as a system, which typically requires a small external power to change parameters of control devices and utilizes the motion of the structure to develop the control force [99].

Semiactive intelligent control devices are usually installed in the deformation position, such as the braces between columns. Control forces are produced through appropriate adjustment of the mechanical properties of the semiactive control devices based on the feedback from sensors that measure the excitation and/or responses of the structure in accordance with a predetermined control algorithm. Like the control modes of active control described in Chapter 3, Active Intelligent Control, the control modes of semiactive control can also be divided into the feedforward control, which is only based on the observation of excitation and the feedback control, which is based on the observation of structural responses, as shown in Fig. 4.1.

4.1.2 Classification

Semiactive control is a kind of structural control technology aiming at parameter control, and it can achieve very good control effect with only a little energy input. In recent years, semiactive control technology develops rapidly and has been widely used in structural vibration control successfully. Till now, the typical semiactive control devices include MR damper, ER damper, piezoelectricity friction damper, semiactive-varied stiffness damper, semiactive-varied damping damper, and MRE device. In the following sections, these semiactive control devices will be introduced in detail, from basic principles, constructions and designs, mathematical models, analysis and design methods, and tests and engineering applications.

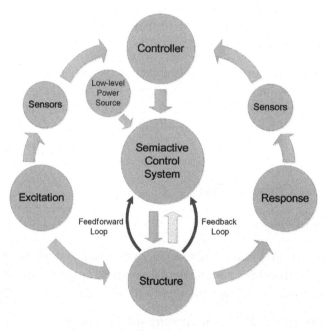

Figure 4.1
Diagram of semiactive control modes.

4.2 MR Dampers

MR damper is designed based on the quick reversible rheological characteristics of MR fluid. The initial discovery and development of MR fluids and devices can be credited to Jacob Rabinow at the US National Bureau of Standards in the late 1940s [100–102]. Due to the complicated design process of magnetic circuit of MR device and early MR fluids have poor stability, there has been scant information published about MR fluids and MR damper before the mid-1980s except for a flurry of interest after their initial discovery. In the recent 20 years, researches on MR fluids and various MR fluid-based devices have regained attention. A number of MR fluids and various MR fluid-based devices have been commercialized in vehicles, machinery and equipment, vibration, and other fields. The main excellent features of MR dampers are that they need very less control power, have large force output and continuous adjustability, quick response to control signal and can be easily combined with other microcomputer controls. Therefore, it has become one of the most important intelligent damping devices for civil engineering structures. It has been preliminary applied in civil engineering structural vibration control and its unique excellent performance and application prospects are obvious.

4.2.1 Basic Principles

MR damper consists of MR fluid, cylinder, piston, and controllable magnetic field. MR fluid typically exhibits rapid, reversible, and tunable transition from Newtonian fluid to a solid-like material under magnetic field. A typical MR fluid consists of 20–40% by volume of relatively pure, 1–7 μm diameter iron particles suspended in a carrier liquid such as mineral oil [103], synthetic oil, water, or glycol. The properties of MR fluid can be controlled in the presence of magnetic field. In the absence of magnetic field, the rheological properties of the MR fluid are similar to that of base fluid except that it is slightly thicker due to the presence of metal particles. In the absence of magnetic field, the metal particles dispersed in the carrier liquid freely and MR fluids exhibit Newtonian-like behavior. However, when a magnetic field is applied, each metal particle becomes a dipole

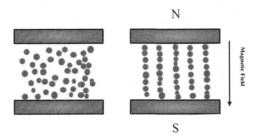

Figure 4.2
Rheological mechanism of MR fluid.

aligning itself along the direction of magnetic field (Fig. 4.2). Thus, a chain-like structure is formed along the direction of magnetic flux. The rheological response of MR fluids results from the polarization induced in the suspended particles by application of an external magnetic field. The interaction between the resulting induced dipoles causes the particles to form columnar structures, parallel to the applied field. These chain-like structures restrict the motion of the fluid, thereby increasing the viscous characteristics of the suspension. The mechanical energy needed to yield these chain-like structures increases as the applied field increases, which results in a field dependent yield stress.

According to the stress pattern of MR fluid, MR damper can be classified as valve mode (or flow mode), direct-shear mode, squeeze mode, and magnetic gradient pinch mode, and diagrams of different working modes are shown in Fig. 4.3.

4.2.1.1 Valve mode

In valve mode of MR fluid operation, the fluid flows through the two fixed surfaces and magnetic field is applied perpendicular to the direction of flow. The resistance of the fluid can be controlled by the intensity of magnetic field. This mode of MR fluid technology is used in various types of dampers and shock absorbers, and has vast application in automobile industry.

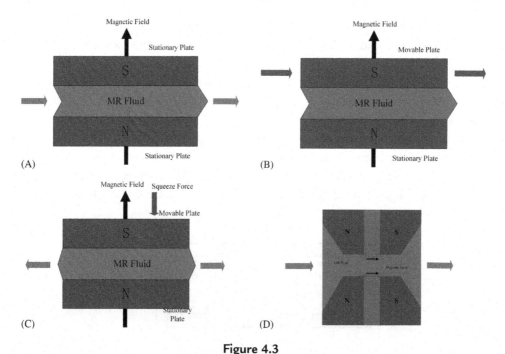

Figure 4.3
Four modes of MR devices operation. (A) Valve mode, (B) direct shear mode, (C) squeeze mode and (D) magnetic gradient pinch mode.

4.2.1.2 Direct-shear mode

In this mode, the fluid flows between surfaces having relative motion and magnetic field is applied perpendicular to the direction of flow. Shear mode of MR fluid technology is used in various types of brakes and clutches of the vehicles.

4.2.1.3 Squeeze mode

In the squeeze mode of operation, MR fluid is placed between two electrodes, which are free to translate in a direction roughly parallel to the direction of the applied magnetic field. Consequently, the MR fluid is subjected to alternate tensile and compressive strokes, and shearing of the fluid occurs at the same time. Available displacements are small but available forces from compact devices are large.

4.2.1.4 Magnetic gradient pinch mode

The latest working mode is called the magnetic gradient pinch mode [104]. The basic idea of this mode is similar to flow mode, but with a different configuration of magnetic circuit design. In the magnetic gradient pinch mode, the magnetic poles are arranged axially along with the flow path and separated by a nonmagnetic material. This kind of pole arrangement will create elliptical magnetic fibrils, which will block the flow of MR fluid in the valve gap. One of the unique characteristics of this mode is that the slope between pressure and velocity relationship in magnetic gradient pinch mode will significantly increase when the magnetic field increases.

Although the MR fluid offers four modes of operation, only the valve mode and the shear mode or a combination of valve and shear modes are used at present in the vast majority of commercial applications, which is mainly on account that the magnetic circuit structure is simple, easy to design. Commercial applications of the squeeze mode MR damper are still restricted to big amplitude vibration damping because of the complex magnetic circuit design and the lack of adequate stroke. Although the magnetic gradient pinch mode has a possibility to use MR fluid with coarser particles, the magnetic gradient pinch mode MR dampers are still under investigated and restricted to commercial applications.

4.2.2 Construction and Design

According to the motions of the pistons, MR dampers can be categorized as linear MR dampers and rotary MR dampers [105]. The principle and configuration of commonly available linear MR dampers based on the valve mode are shown in Fig. 4.4, which is developed by the Lord Corporation [106]. The cylinder is divided into upper and lower chambers by the piston and the MR fluid flows through the annular gap between the outside circumference of the piston and the inside wall of the cylinder. The magnetic circuit is formulated along with the piston and cylinder, and the MR fluid in the fluid

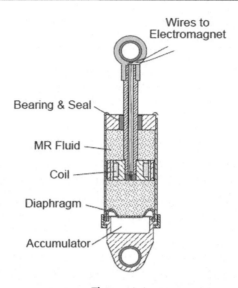

Figure 4.4
Schematic and configuration of an MR damper operated in the valve mode.

gap will be exposed to the magnetic field generated by the exciting coil wound on the piston. According to the MR effect, the damping force of MR damper can be controlled under the external current to some extent, and the current driver applies the current to the exciting coil according to the command voltage determined by the corresponding controller.

The schematic of commonly available rotary MR damper operated in the direct-shear mode is shown in Fig. 4.5. In this damper, MR fluid is located between the faces of the disc-shaped rotor and the stationary housing. Rotation of the shaft causes the MR fluid to be directly sheared as the rotor moves relative to the housing. A coil fixed in the housing produces a toroid-shaped magnetic field that interacts with the MR fluid in the fluid gaps on each side of the rotor. Shear mode MR dampers are similar to conventional passive shear mode dampers, the former can produce additional damping force or torque resulting from the yield stress of MR fluids in response to the magnetic field. This section will focus primarily on the design method of linear MR dampers.

A valve mode or a mix mode MR damper in axi-symmetric case, as well as the approximated parallel plate case, has been well analyzed by Kamath et al. [107]. The results show that the error in approximating an axi-symmetric valve with parallel plates is small when the ratio between flow gap h and diameter D is far less than one ($h/D<<1$), so that many MR dampers can be modeled by the parallel plates approximation. Therefore, for simplicity, the parallel-plate model is readily used in the initial design of the MR dampers.

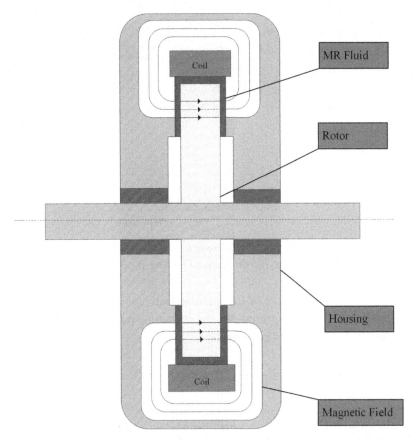

Figure 4.5
Schematic of an MR damper operated in the direct-shear mode.

In order to design a MR damper with the operation of valve mode, Mao et al. [108] proposed a MR damper model based on parallel-plate model. The schematic diagram of annular duct type MR damper is presented in Fig. 4.6. According to the proposed MR damper model, the damping force F of the MR damper can be expressed as follows:

$$F = F_\eta + F_\tau = A_p \left(f \frac{\rho L \overline{A}^2 V_p^2}{4h} + \frac{2L\tau_y}{h} \right) \quad (4.1)$$

where F_η is the damper force due to the fluid viscosity, F_τ is the damper force due to the shear yield stress under a magnetic field, L is the effective length of the flow channel in piston and $L = L_1 + L_2 + L_3$, A_p is the effective piston area and $A_p = \pi(D-d)^2/4$, A_d is the cross-sectional area of the gap and $A_d = h \cdot \pi D$, the dimensionless number $\overline{A} = A_p/A_d$, D is the effective gap diameter, d is the diameter of the piston rod, ρ is the density of MR fluid,

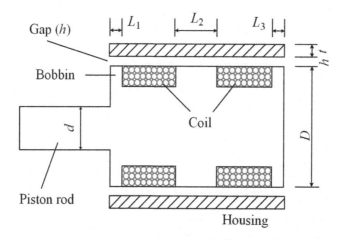

Figure 4.6
MR fluid path in the piston head of a valve mode MR damper.

V_p is the piston velocity, h is the size of the annular flow gap, f is the Darcy friction factor and is chosen depending on the magnitude of the Reynolds number of the fluid flow, and the friction factor for parallel plates is given by

$$f = \frac{96}{\text{Re}_{D_h}} \quad \text{if} \quad \text{Re}_{D_h} \leq 2300 \tag{4.2}$$

$$\frac{1}{f^{1/2}} = -1.8 \log_{10}\left[\left(\frac{\varepsilon/D_h}{3.7}\right)^{1.11} + \frac{6.9}{\text{Re}_{D_h}}\right] \quad \text{if} \quad \text{Re}_{D_h} \geq 4000 \tag{4.3}$$

where ε is the average pipe housing roughness and assumed to be 0.01 mm in this study, D_h is the hydraulic diameter, and $D_h = 2h$, Re_{D_h} is the Reynolds number and defined by $\text{Re}_{D_h} = 2\rho \bar{A} V_p h / \eta$, η is the apparent viscosity of MR fluid in the absence of a magnetic field.

The dynamic range λ_d is defined as the ratio of the total damper force to the uncontrollable force as follows:

$$\lambda_d = \frac{F_\eta + F_\tau}{F_\eta} = 1 + \frac{8\tau_y}{f \rho \bar{A}^2 V_p^2} \tag{4.4}$$

By defining Bingham number as follows:

$$Bi = \frac{\tau_y h}{\eta \bar{A} V_p} \tag{4.5}$$

The dynamic ranges are as follows:

$$\lambda_d = 1 + \frac{Bi}{6} \quad \text{if } \text{Re}_{D_h} \leq 2300 \tag{4.6}$$

$$\lambda_d = 1 + \frac{51.84 \, Bi \left(\log_{10} \left[\left(\frac{\varepsilon/D_h}{3.7} \right)^{1.11} + \frac{6.9}{\text{Re}_{D_h}} \right] \right)^2}{\text{Re}_{D_h}} \quad \text{if } \text{Re}_{D_h} \geq 4000 \tag{4.7}$$

Then, based on the proposed model, an effective MR damper design procedure is also proposed:

1. Specify the desired dynamic range, λ_d, and the zero-field damping force, F_0, at the maximum piston velocity of interest, V_p.
2. Specify fluid properties of MR fluid such as τ_y, ρ, η.
3. Calculate Bingham number form $Bi = 6(\lambda_d - 1)$.
4. Determine an appropriate h, $h = \sqrt{\eta^2 \cdot Bi \cdot \text{Re}/2\rho\tau_y}$, so that the Reynolds number $\text{Re}_{D_h} \ll 4000$ (Smaller Reynolds number is recommended for a useful dynamic range).
5. Determine $\bar{A}$ by putting h determined from step 4 into $\bar{A} = \tau_y h/\eta V_p Bi$.
6. Then, depending on design restrictions to be primarily considered, follow anyone of the cases below:
    i. If the effective length L is critically constrained, determine A_p from $A_p = 4hF_0/f(\rho\bar{A}^2 V_p^2)L$ using the allowable L, then compute A_p from $\bar{A}$ obtained from step 5.
    ii. If the inner-diameter size of the MR damper is critically constrained, namely, A_p is restricted, then determine L from $L = 4hF_0/f(\rho\bar{A}^2 V_p^2)A_p$ using allowable A_p.
    iii. If there are no such space limitations, choose any A_p and A_d producing the above $\bar{A}$ and calculate L from the same equation in (ii).
7. Calculate the effective diameter of the fluid path D from $D = A_d/(\pi d)$.
8. Based on the above information, given a range of interested $\text{Re}(\text{Re} \ll 4000)$, plots the corresponding figures, and chooses the appropriate design parameter values from these figures.

Finally, the proposed model and the design strategy were validated by mechanical test on the semiactive MR damper [108], and the cross-sectional view of the final designed MR damper is presented in Fig. 4.7, and the parameters are: $\text{Re} = 195$, $h = 0.6$ mm, $\bar{A} = 10.5$, $L = 44$ mm, $A_d = 5.56 \times 10^{-5}$ m^2, $D = 29.43$ mm, respectively.

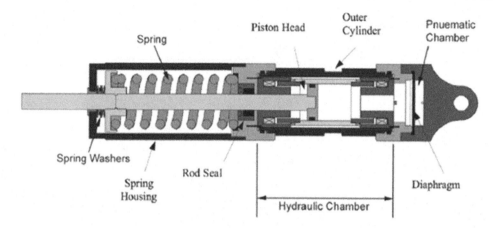

Figure 4.7
Cross-sectional view of MR damper with spring mechanisms.

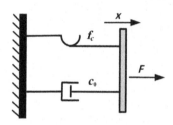

Figure 4.8
Bingham model.

4.2.3 Mathematical Models

How to establish a mathematical model to describe the complex nonlinearity of MR damper precisely is a challenging topic. Up to now, some mathematical models have been proposed, such as Bingham model and modified Bingham model, nonlinear hysteretic biviscous model, Bouc–Wen hysteresis dynamic model, and so on. These models will be introduced in the following.

4.2.3.1 Bingham model and modified Bingham model

This model is the most simplified model used to calculate the damping force for MR dampers, which is composed of a friction element and a viscous element. According to Fig. 4.8, the force generated by MR damper is given by [109]

$$F(t) = c_0 \dot{x} + f_c \operatorname{sgn}(\dot{x}) + f_0 \tag{4.8}$$

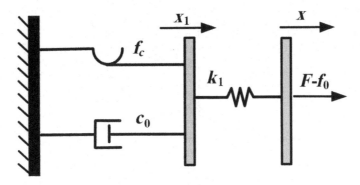

Figure 4.9
Modified Bingham model.

where $\dot{x}$ denotes the velocity attributed to the external excitation, c_0 is the damping coefficient, f_c is the frictional force related to the field-dependent yield stress and f_0 is the offset in the force included to account for nonzero mean observed in the measured force due to the presence of the accumulator.

The modified Bingham model is shown in Fig. 4.9, in which a spring member is connected in series with the Bingham model. The governing equation is as follows [110]:

$$F(t) = c_0 \dot{x}_1 + f_c \, \text{sgn}(\dot{x}_1) + f_0 = k_1(x - x_1) + f_0 \tag{4.9}$$

where k_1 is the equivalent axial stiffness of the MR damper, which can be considered as the stiffness of the accumulator and the initial shear modulus of the MR fluid in the pre-yield phase, the other parameters are the same as Eq. (4.8).

4.2.3.2 Nonlinear hysteretic biviscous model

Observing the damper behavior during testing, the force–velocity hysteretic curve shows a distinct preyield hysteresis. A four-parameter nonlinear hysteretic biviscous model was proposed, as shown in Fig. 4.10, and equations of the piecewise continuous nonlinear hysteretic biviscous model are as follows [111,112].

$$F(t) = \begin{cases} C_{\text{post}} \dot{x} - F_y & \dot{x} \leq -\dot{x}_{y1} & \ddot{x} > 0 \\ C_{\text{pre}}(\dot{x} - \dot{x}_0) & -\dot{x}_{y1} \leq \dot{x} \leq \dot{x}_{y2} & \ddot{x} > 0 \\ C_{\text{post}} \dot{x} + F_y & \dot{x}_{y2} \leq \dot{x} & \ddot{x} > 0 \\ C_{\text{post}} \dot{x} + F_y & \dot{x}_{y1} \leq \dot{x} & \ddot{x} < 0 \\ C_{\text{pre}}(\dot{x} + \dot{x}_0) & -\dot{x}_{y2} \leq \dot{x} \leq \dot{x}_{y1} & \ddot{x} < 0 \\ C_{\text{post}} \dot{x} - F_y & \dot{x} \leq -\dot{x}_{y2} & \ddot{x} < 0 \end{cases} \tag{4.10}$$

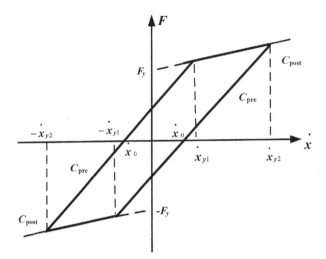

Figure 4.10
Nonlinear hysteretic biviscous dynamic model for MR dampers.

where C_{pre} and C_{post} are the preyield and postyield damping coefficients, respectively, F_y is the yield force represented by the postyield force versus velocity asymptote intercept with the force axis, $\dot{x}_0$ is the zero force−velocity intercept, and $\dot{x}_{y1}$ and $\dot{x}_{y2}$ are the compressive (decelerating) and tensile (accelerating) yield velocities, which are given by

$$\dot{x}_{y1} = \frac{F_y - C_{pre}\dot{x}_0}{C_{pre} - C_{post}} \quad \text{and} \quad \dot{x}_{y2} = \frac{F_y + C_{pre}\dot{x}_0}{C_{pre} - C_{post}} \quad (4.11)$$

The hysteresis cycle is separated into two groups of equations. Among the expression of Eq. (4.10), the first three are for positive acceleration, while the last three are for negative acceleration. In order to accurately characterize the behavior of MR dampers using the nonlinear hysteretic biviscous model, a set of four constant parameters $\Theta = [C_{pre}, C_{post}, F_y, \dot{x}_0]$ should be identified.

4.2.3.3 Bouc−Wen hysteresis model

The Bouc−Wen hysteresis operator was initially formulated by Bouc [113] as an analytical description of a smooth hysteretic model and later generalized by Wen [114]. The Bouc−Wen model has been extensively applied to simulate the hysteresis loops since it possesses the force−displacement and force−velocity behavior of the MR damper. Spencer et al. [109] adopted the Bouc−Wen hysteretic operator to represent the hysteretic behavior of MR dampers and the schematic of the proposed simple Bouc−Wen model for MR dampers is shown in Fig. 4.11. The damping force in this system is given by

$$F(t) = c_0\dot{x} + k_0(x - x_0) + \alpha z \quad (4.12)$$

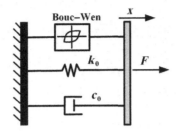

Figure 4.11
Bouc−Wen model for MR dampers.

where c_0 and k_0 are the viscous and stiffness coefficients, respectively, an initial displacement x_0 of the spring is incorporated into the model to allow for the presence of an accumulator in the considered damper, and z is an evolutionary variable governed by

$$\dot{z} = -\gamma |\dot{x}| z |z|^{n-1} - \beta \dot{x} |z|^n + A\dot{x} \tag{4.13}$$

where $\dot{z}$ denotes the time derivative. The scale and general shape of the hysteresis loop are governed by A, γ, and β, while the smoothness of the force−displacement curve is controlled by n. The whole parameters are $\Theta = [c_0, k_0, \alpha, x_0, \gamma, \beta, A, n]$, which can be identified using the experimental data.

4.2.3.4 Dahl model and modified Dahl model

The Dahl model [115] was developed for the purpose of simulating control systems with friction. The Dahl model can be written as

$$\dot{z} = \frac{\sigma}{f_c}(\dot{x} - |\dot{x}|z) \tag{4.14}$$

$$F(t) = f_c z \tag{4.15}$$

in which $z(t) = (\sigma/f_c)Z(t)$, where σ is the stiffness coefficient, Z is the internal hysteretic variable, which is governed by a nonlinear filter, and f_c is the Coulomb friction force.

Zhou [110] suggested a simple and effective modified Dahl model for MR dampers as shown in Fig. 4.12. The damper force is given by

$$F(t) = k_0 x(t) + c_0 \dot{x} + \alpha z - f_0 \tag{4.16}$$

where z is a nondimensional hysteretic variable governed by Eq. (4.14). In order to calibrate the modified Dahl model under an applied fluctuating magnetic field, it is necessary to obtain the relationship between model parameters and applied magnetic field. The experimental results show that α and c_0 are related to the applied voltage determined by the following equations, respectively.

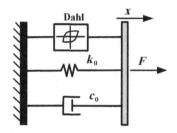

Figure 4.12
Modified Dahl model for MR dampers.

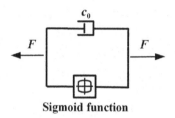

Figure 4.13
Sigmoid model.

$$\alpha = \alpha(u) = \alpha_a + \alpha_b u \qquad (4.17)$$

$$c_0 = c_0(u) = c_{0a} + c_{0b} u \qquad (4.18)$$

where c_{0a} and α_a are the damping coefficient and the Coulomb force of the MR damper at 0 V, respectively, and u is given as the first-order filter given by

$$\dot{u} = -\eta(u - v) \qquad (4.19)$$

where η reflects the response time of the MR damper, namely, larger η means faster response time, and v is the command voltage sent to the current driver.

In the modified Dahl model, the Dahl hysteresis operator instead of the Bouc–Wen hysteresis operator is adopted to simulate the Coulomb force to avoid the determination of many parameters and the modified Dahl model can well capture the force–velocity relationship in the low-velocity region. A set of eight parameters should be determined for the proposed modified Dahl model for MR dampers and the set of parameters is $\Theta = [c_{0a}, c_{0b}, \alpha_a, \alpha_b, k_0, \sigma, f_0, n]$.

4.2.3.5 Sigmoid model

Xu and Shen [116] considered that the typical force–velocity characteristics of an MR damper can be represented by the Sigmoid function, i.e., Sigmoid model, as shown in Fig. 4.13, the damping force can be given by

$$F(t) = F_y \frac{1 - e^{\beta \dot{x}/\omega}}{1 + e^{\beta \dot{x}/\omega}} + c_0 \dot{x} \qquad (4.20)$$

where F_y is the yield force, $\beta > 0$ is the index, c_0 and ω are the viscous and circular frequencies.

In order to accurately characterize the behavior of MR dampers using the Sigmoid model for MR dampers, a set of three constant parameters $\Theta = [c_0, F_y, \beta]$ should be identified using the experimental data.

4.2.3.6 Magnetic saturation mathematical model

In order to consider the magnetic saturation effect of MR dampers, Xu et al. [117] proposed the magnetic saturation model, which is simplified as a viscous damper element in parallel with tanh stiffness element and spring element and then in series with mass element, as shown in Fig. 4.14. The damping force is given by

$$F = m\ddot{u} + C_c \dot{u} + K_k u + F_y \tanh(\beta(\dot{u} + \lambda \operatorname{sgn}(u))) + F_0 \qquad (4.21)$$

where F and u are the damping force and displacement of the MR damper; C_c is the variable damping coefficient; K_k is the variable stiffness coefficient; m is the equivalent mass that represents the inertial effect of the piston rod; F_y is the damping force related to magnetic fields; F_0 is the initial force related to initial position; β and λ are shape factors on the smooth degree of the hysteretic curves.

4.2.4 Analysis and Design Methods

When MR dampers are incorporated in a structure, as shown in Fig. 4.15, the equation of motion of the controlled structure can be expressed as Eq. (2.1).

In the analysis of the structure with MR dampers, the most important is to determine the control force vector $f_d(t)$ of MR dampers, the Linear Quadratic Regulator (LQR) control

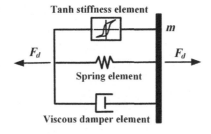

Figure 4.14
Equivalent model for MR dampers.

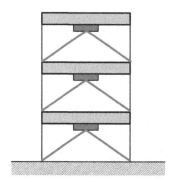

Figure 4.15
Diagram of the controlled structure.

algorithm described in Chapter 2, Intelligent Control Strategies, is used here. The instantaneous quadratic objective function can be chosen as

$$J(t) = \{z(t)\}^T[Q]\{z(t)\} + \{f_d(t)\}^T[R]\{f_d(t)\} \tag{4.22}$$

where $\{z(t)\}$ is the state vector, $\{f_d(t)\}$ is the control force vector, $[Q]$ and $[R]$ are weight matrices to adjust the importance of structural response and control force.

To minimize the quadratic objective function, the optimal control force can be obtained as

$$\{f_d(t)\} = \left(-\frac{\Delta t}{2}\right)[R]^{-1}[B]^T[Q]\{z(t)\} \tag{4.23}$$

where Δt is sampling period, $[B]$ is the state matrix of the system.

However, MR damper is a semiactive control device by changing the magnetic field to adjust the control force, it cannot achieve the optimal control force calculated by the above equation in any instantaneous, and thus a semiactive control strategy should be adopted to deal with the problem.

Here, the idea of multistate control is taken for example to adjust the inputting current. The main idea is as follows: first, the coulomb damping force is calculated at the given level of current according to the classic quasi-static model. Second, the result is compared with the former calculated optimal control force, and the coulomb damping force with a minimum difference is selected as the intelligent control force of the MR damper. If the coulomb damping force is in the opposite direction to the optimal control force, the intelligent control force is set as zero. The corresponding multistate control strategy is shown in Fig. 4.16.

Of course, some other optimal control algorithms and semiactive control strategies can also be used according to the actual problems. Based on the equation of motion and semiactive control algorithm, the semiactive control can be realized and the dynamic responses of the controlled structure incorporated with MR dampers can be calculated.

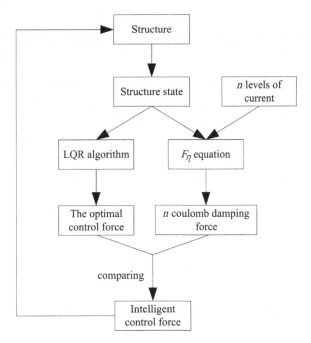

Figure 4.16
Multistate control strategy.

4.2.5 Tests and Engineering Applications

To prove the scalability of MR fluid technology to devices of the appropriate size for civil engineering applications, Dyke et al. [106] carried out the performance test on the large-scale 200 kN MR fluid damper. Sanwa Tekki Cooperation used the MR fluid developed by Lord company to develop dampers with the largest damping force at that time, the damping force tested are 300 and 400 kN, respectively [118], and the 400 kN MR damper was installed in a residential building at Keio University in Japan; Ou and Guan [119] designed a mixed mode MR damper, the maximum damping force is 2.7 kN. Li et al. [120] also designed and fabricated a mixed mode MR damper, performance test was conducted and the maximum force is 21.3 kN. Moon et al. [121] designed and fabricated an MR damper, which has the capacity of about 10 kN, and a series of tests have been performed to grasp the fundamental performance characteristics of the MR damper. Qu et al. [122] designed and fabricated a 500 kN large-scale MR damper, and performance test on this damper was also carried out. Oh and Onoda experimentally studied the semi-active magnetorheological fluid variable damper for vibration suppression of truss structures.

In 2006, Xu et al. [117] designed and prepared a shear-valve mode MR damper with the maximum force of 200 kN by using self-made MR fluid, the schematic and main geometric parameters of this damper can be seen in Fig. 4.17 and Table 4.1, respectively.

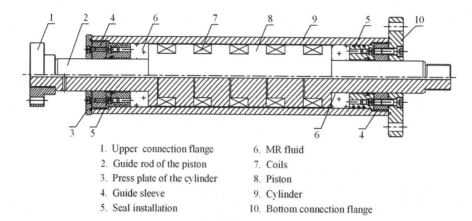

1. Upper connection flange
2. Guide rod of the piston
3. Press plate of the cylinder
4. Guide sleeve
5. Seal installation
6. MR fluid
7. Coils
8. Piston
9. Cylinder
10. Bottom connection flange

Figure 4.17
Schematic of MR damper.

Table 4.1: Main geometric parameters of MR damper

Stroke (mm)	±50
External diameter of cylinder (mm)	194
Internal diameter of cylinder (mm)	160
Diameter of the piston rod (mm)	80
Effective length of the piston (mm)	4 × 50 + 2 × 25 = 250
Damping gap (mm)	2
Trench diameter of coil (mm)	110

In order to evaluate the properties of the manufactured MR damper, including mechanical behavior and energy dissipation performance, performance tests on this damper were carried out at different input current, displacement amplitude, and frequency, as shown in Fig. 4.18. In testing processes, the actuator of the loading device is controlled by input displacement with a sinusoidal waveform $u = A \sin(2\pi ft)$, where A is the displacement amplitude: 10, 20, 30, and 40 mm. f is the loading frequency: 0.1, 0.2, 0.5, and 1.0 Hz. The input currents are 0, 0.6, 0.9, 1.2, 1.5, 1.8, 2.4, and 2.7 A, respectively. t is the loading time. Taking the condition of five coils is energized, for example, the test results will be discussed in detail.

Fig. 4.19 shows the force−displacement and force−velocity hysteresis curves variation with different input currents, in which the displacement amplitude is 20 mm and the loading frequency is 0.2 Hz. The input currents in every coil are 0, 0.6, 0.9, 1.2, 1.5, 1.8, 2.4, and 2.7 A, respectively.

Figure 4.18
Performance test on MR damper.

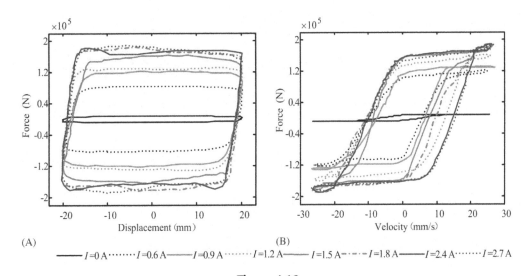

Figure 4.19
Hysteresis curves of MR damper when five coils are energized. (A) Force–displacement hysteretic curves. (B) Force–velocity hysteretic curves.

As seen in Fig. 4.19, when five coils are energized from 0 to 1.8 A, the amplitudes of the generated damping force are 9.42, 83.77, 123.40, 132.27, 163.96, and 184.06 kN, respectively, the corresponding increments of damping forces are 74.35, 39.63, 8.87, 31.69, and 20.10 kN, respectively, the damping force and the area of hysteresis curves increase greatly. However, when five coils are energized from 1.8 to 2.4 A and 2.7 A, the amplitudes of the generated damping force are 184.06, 186.20, and 188.91 kN, respectively, the corresponding increments of damping forces are 2.14 and 2.71 kN, respectively, which are very small as compared to previous increments. In addition, the maximum and minimum damping force of the MR damper are 188.91 and 9.42 kN, respectively, then the dynamic range of the MR damper at the displacement amplitude of 20 mm and the loading frequency of 0.2 Hz can be calculated in accordance with the equation, $K = F_\tau / F_{uc} = (188.91 - 9.42)/9.42 \approx 19$, where F_τ is the controllable damping force produced by magnetic field at different input current, and F_{uc} is the uncontrollable force.

It is also obvious that the damping force of the MR damper significantly increases with the input currents change from 0 to 1.8 A. However, the damping forces tend to be stable during the scope from 1.8 to 2.4 A. Therefore, the existence of magnetic saturation is verified by the tests, and the current range of the magnetic saturation is about 1.8 to 2.4 A.

Fig. 4.20 shows the variation of force−displacement and force−velocity hysteresis curves with displacement amplitude when five coils are energized, in which the loading frequency is 0.2 Hz, and the current are 1.8, 2.4, and 2.7 A. It can be seen that the maximum damping force increases slightly with the increase of displacement amplitude when the loading

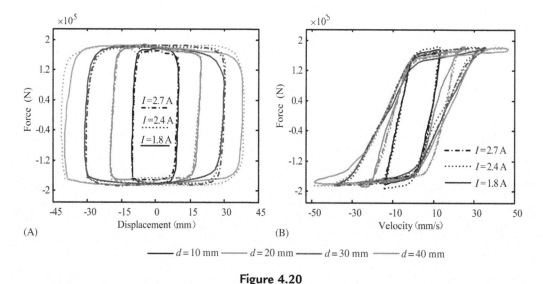

Figure 4.20
Hysteresis curves with the displacement amplitude when five coils are energized. (A) Force−displacement hysteretic curves. (B) Force−velocity hysteretic curves.

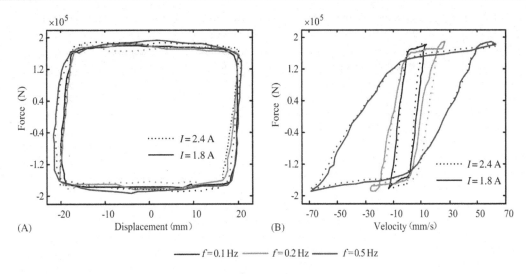

Figure 4.21
Hysteresis curves with the loading frequency in the magnetic saturation situation when five coils are energized. (A) Force–displacement hysteretic curves. (B) Force–velocity hysteretic curves.

frequency and the input current are given. However, the area surrounded by hysteresis curve, i.e., the energy dissipation of the damper increases greatly with the increase of displacement amplitude.

Fig. 4.21 shows the variation of force–displacement and force–velocity hysteresis curves with loading frequency when five coils are energized, in which the displacement amplitude is 20 mm, the current are 1.8 and 2.4 A. It can be seen that the maximum damping force increases slightly with the increase of loading frequency when the displacement amplitude and the input current are given. The force–displacement hysteresis curves nearly overlap. These phenomena demonstrate that the effect of loading frequency on the damping force and the energy dissipation is negligibly small.

Generally, energy dissipation capacity of the MR damper E_d can be obtained by calculating the area enclosed by the force–displacement hysteresis curve, E_d is given by

$$E_d = \int F(t)du = \int_0^{2\pi/\omega} F(t)\dot{u}(t)dt \qquad (4.24)$$

where $\dot{u}(t)$ is the velocity of the piston, ω is the loading circular frequency.

Table 4.2 shows the energy dissipation values when five coils are energized from 0 to 2.4 A at the interval of 0.6 A. When the loading frequency is 0.2 Hz and the displacement amplitude is 30 mm, the energy dissipation values under different input currents are 0.86, 8.80, 13.26, 17.23, and 17.67 kJ, respectively, and the relative increments between two

Table 4.2: Energy dissipation value of MR damper when five coils are energized from 0 to 2.4 A

Frequency	Displacement Amplitude	Current				
		$I = 0$ A	$I = 0.6$ A	$I = 1.2$ A	$I = 1.8$ A	$I = 2.4$ A
$f = 0.1$ Hz	$A = 10$ mm	0.65	2.05	2.62	2.83	3.44
	$A = 20$ mm	1.47	5.48	7.71	7.54	10.48
	$A = 30$ mm	2.47	8.67	12.89	12.93	17.44
	$A = 40$ mm	3.54	7.56	18.06	17.98	25.32
$f = 0.2$ Hz	$A = 10$ mm	0.25	2.01	2.83	3.56	3.51
	$A = 20$ mm	0.51	4.32	8.09	10.36	10.14
	$A = 30$ mm	0.86	8.80	13.26	17.23	17.67
	$A = 40$ mm	1.16	11.22	17.19	22.60	23.92
$f = 0.5$ Hz	$A = 10$ mm	0.27	2.10	2.98	3.51	3.35
	$A = 20$ mm	0.68	5.76	8.60	11.24	10.88
	$A = 30$ mm	1.26	9.38	13.83	15.55	18.78
	$A = 40$ mm	1.62	13.57	19.91	22.57	21.97

adjacent input currents are correspondingly 923.26%, 50.68%, 29.94%, and 2.55%. It is clearly observed that the increment of the energy dissipation is very small when five coils are energized from 1.8 to 2.4 A. Therefore, the existence of magnetic saturation is verified by these phenomena once more, at the same time, the magnetic saturation of the current range of 1.8 to 2.4 A is determined once again.

In order to investigate the effects of input current, displacement amplitude, and loading frequency on the energy dissipation in the magnetic saturation situation, the variation of energy dissipation with input current, displacement amplitude, and loading frequency are plotted in Figs. 4.22–4.24, respectively.

Fig. 4.22 shows the variation of energy dissipation with the input current. It can be seen that the energy dissipation of the damper increases significantly as the input current increases when the input current is less than 1.8 A. However, the energy dissipation capacity is nearly unchanged when the input current increases from 1.8 to 2.4 A. These phenomena demonstrate that the effect of the input current on energy dissipation capacity is negligibly small under the magnetic saturation situation.

Fig. 4.23 shows the variation of energy dissipation with the displacement amplitude, in which the loading frequency is 0.2 Hz. It can be seen that the energy dissipation of the damper increases significantly with the increase of displacement amplitude when the input current and loading frequency are given.

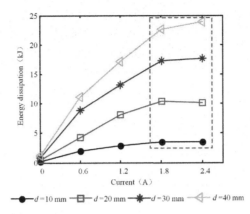

Figure 4.22
The variation of energy dissipation with the current.

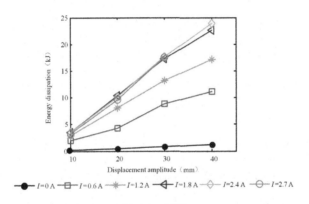

Figure 4.23
The variation of energy dissipation with the displacement amplitude.

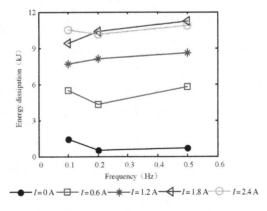

Figure 4.24
The variation of energy dissipation with the frequency.

Fig. 4.24 shows the variation of energy dissipation with the loading frequency. It can be seen that the alteration of the damper energy dissipation is not obvious with the change of loading frequency when the input current and the amplitude are given. These phenomena demonstrate that the effect of loading frequency on energy dissipation capacity is negligibly small.

In recent years, shaking table tests of structures incorporated with MR dampers were also conducted. Li et al. [124] carried out the shaking table test and experimentally studied the feasibility of nonlinear seismic control of a three-story steel-concrete hybrid structure using MR dampers. Li et al. [125] also carried out a test to study the stochastic seismic response control of structures with MR dampers.

Due to the excellent properties, MR dampers have been used in some practical engineering projects. MR dampers can provide controllable forces and offer unique solutions for the control of dynamic response of buildings or bridges. The first full-scale implementation of MR dampers for civil engineering applications was achieved in 2001 [126]. Two 300 kN MR dampers were installed between the third and fifth floors in the Tokyo National Museum of Emerging Science and Innovation to control the vibration due to seismic excitation, as shown in Fig. 4.25.

Retrofitted with stay-cable dampers, the Dongting Lake Bridge in Hunan, China constitutes the first full-scale implementation of MR dampers for bridge structures, shown in Fig. 4.26. Two Lord SD-1005 MR dampers are mounted on each cable to mitigate cable vibration. A total of 312 MR dampers are installed on 156 stayed cable. The technical support for this engineering project was provided through a joint venture between Central South University

Figure 4.25
Nihon-Kagaku-Miraikan, Tokyo National Museum of Emerging Science and Innovation, installed with 30 t MR fluid dampers.

Figure 4.26
MR damper installation on the Dongting Lake Bridge, Hunan, China.

Figure 4.27
MR damper installation on the Xi'erhuan Bridge, Hanzhong, China.

and The Hong Kong Polytechnic University [127]. After that, Ou used MR dampers to reduce the wind-rain induced vibration of stay-cables in the Binzhou Yellow River Bridge and Ningbo Zhaobaoshan Bridge in China [27]. Fujitani [128] used MR dampers in a real-isolated building with a maximum damping force of 400 kN. Qu [129,130] used a large-scale 500 kN MR damper to reduce the braking-induced longitudinal vibration responses of the Wuhan Tianxinzhou Yangtz River Bridge.

Xu and Wang used 44 MR dampers to control the vibration of cables of Xi'erhuan Bridge in Hanzhong city, as shown in Fig. 4.27. Nowadays, Xu is trying to use MR dampers to control the wind-induced vibration of cables of the Nanjing Yangtze river second bridge, two MR dampers are installed on the longest cable and the solar panel is employed to acquire energy to provide power for the MR dampers, and the controller gathering sensing, controlling and storage capacity together is adopted to realize semiactive control of cables' vibration due to strong wind.

4.3 ER Dampers

ER damper is designed based on the rheological characteristics of ER fluid. ER fluid is a type of "intelligent" colloid capable of varying viscosity or even solidification in response to an applied electric field. The ER phenomenon of ER fluid was discovered more than 60 years ago, widely credited to Winslow [131]. The rheological variation is reversible when the field is removed. The response time can occur within 10 ms. Due to such amazing features, ER fluids can serve as an electric–mechanical interface, and when they are coupled with sensors (as triggers to activate the electric field), many mechanical devices such as clutches, valves, and dampers may be converted into active mechanical elements, which are capable of responding to environmental variations.

The diverse applications potential [97–99,131–134] has made ER fluids a persistent area of study in soft matter research, ever since their discovery six decades ago. However, in spite of the broad interest, applications have been hampered by the weakness of the ER effect. This state of affairs has been changed in recent years owing to the discovery of the giant ER effect [135,136], which represents a different paradigm from the conventional ER mechanism. The discovery of the giant ER also facilitated the application of ER fluids in various devices.

In this chapter, ER dampers will be introduced in detail from the aspect of basic principles, construction and design, mathematical models, analysis and design methods, and tests and engineering applications.

4.3.1 Basic Principles

The principle of ER damper is very similar to that of MR damper, the classical ER damper also consists of ER fluid, cylinder, piston, and controllable electric field. ER fluid typically exhibits rapid, reversible, and tunable transition from Newtonian fluid to a solid-like material under an electric field. ER damper, which takes the advantages of ER fluids, is a promising semiactive control device.

A description of the ER effect of the ER fluid is as follows. When the electric field is absence, ER fluids exhibit near-Newtonian fluids properties, like normal oils. When an electric field $\vec{E}$ is applied to the fluid, owing to the dielectric constant contrast between the solid particles and the liquid in a colloid, each solid particle would be polarized under an electrostatic field, with an effective dipole moment. If ε_s denotes the complex dielectric constant of the solid particles and ε_l denotes the complex dielectric constant of the liquid, R is the radius of particle, then the induced dipole moment can be expressed as follows

$$\vec{p} = \frac{\varepsilon_s - \varepsilon_l}{\varepsilon_s + 2\varepsilon_l} R^3 \vec{E}_l = \beta R^3 \vec{E}_l \tag{4.25}$$

where β is the Claussius-Mossotti factor, and $\vec{E}_l$ should be understood as the field at the location of the particle, i.e., the local field. The resulting (induced) dipole–dipole interaction between the particles means that the random dispersion is not the lowest energy state of the system, and particles would tend to aggregate and form columns along the applied field direction. The formation of columns is the reason why the high-field state of ER fluid exhibits increased viscosity or even solid-like behavior, able to sustain shear in the direction perpendicular to the applied electric field [134,135]. This picture is schematically illustrated in Fig. 4.28.

However, the formation of the chains/columns is governed by the competition between electrical energy and the Brownian motion of the particles, which can be manifested by using the value of the dimensionless parameter $\gamma = \vec{p} \cdot \vec{E}/k_B T$, where k_B is the Boltzmann constant and T is the temperature (the room temperature can be employed), p is given by Eq. (4.25). The chain will form when $\gamma > 1$. Thus for $\beta^{1/3} R \sim 100$ nm, the field should be larger than 2 kV cm^{-1} for the system to form the chain.

Electric polarization of a collection of atoms/molecules can arise in two ways from an applied electric field. One is the induced polarization process that underlies the traditional ER effect, described above. This type is denoted as the dielectric electrorheological (DER) effect. The other is by aligning the initially randomly oriented molecular dipoles. This second process is responsible for the DER effect.

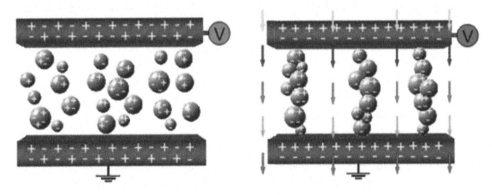

Figure 4.28
ER behavior of ER fluid.

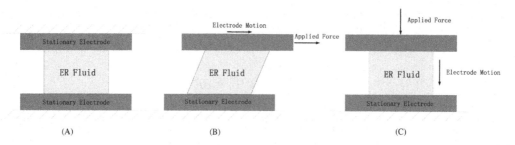

Figure 4.29
Modes of operation of smart fluids. (A) flow mode, (B) shear mode, and (C) squeeze-flow mode.

There are three key modes of operation for ER fluids in diverse ER devices [136], which are illustrated in Fig. 4.29. Fig. 4.29A shows the flow mode of operation. In the flow mode, the ER fluid is contained between a pair of stationary electrodes. The resistance to the fluid flow is controlled by varying the strength of the electric field across the electrodes, in other word, the modulating action of the applied electric field can be used to control the pressure drop or volume flow rate characteristics of the ER fluid around the hydraulic circuit.

The second mode of operation is the shear mode (Fig. 4.29B), which allows relative motion, either rotational or translational, roughly at right angles to the direction of the applied electric field, and the ER fluid is subjected to shear force at this time. Since the shear stress or shear rate characteristics can be controlled continuously through the applied field, this arrangement forms the basis of a variety of engineering devices. Controllable brakes and clutches can be constructed by exploiting the shear mode of operation and relatively simple vibration dampers (both translational and rotational) can be built based on this operational mode.

The third possible mode of operation, named squeeze-flow mode, is illustrated in Fig. 4.29C. Here, the electrodes are free to translate in a direction roughly parallel to the direction of the applied electric field. In squeeze-flow mode, the ER fluid is subjected to alternate tension/compression and some shearing of the fluid also occurs. The displacements available in squeeze-flow are only the order of a few millimeters, but large forces are available from compact devices and there are many potential applications, notably in vibration isolation.

4.3.2 Construction and Design

The schematic configuration of a typical linear ER damper is shown in Fig. 4.30. The ER damper is composed of the electrode, piston, and gas chamber. The diaphragm between the cylinder and the gas chamber is also used in order to compensate for the volume induced by the motion of the piston. The electrode and the gas chamber in the ER damper are fully filled with the ER fluid and nitrogen gas, respectively. The ER damper is divided into upper and lower

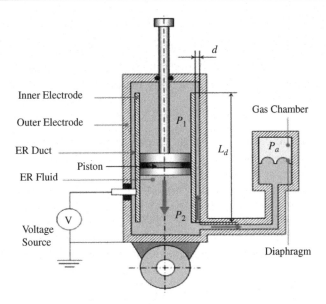

Figure 4.30
Schematic and configuration of the linear ER damper operated in the valve mode.

chambers by the piston, and the ER fluid flows through the annular duct between the inner and outer electrodes from one chamber to the other by the motion of the piston. Thus, the operating mode of the proposed ER damper is flow mode in which two electrodes are fixed. The control voltage generated by a high-voltage supply unit is connected to the inner electrode and the ground voltage is connected to the outer electrode. In the absence of the electric field, the ER damper produces a damping force caused only by fluid viscous resistance. However, if a certain level of the electric field is supplied to the ER damper, the ER damper produces an additional damping force owing to the yield stress of the ER fluid. The damping force of the ER damper can be continuously tuned by controlling the intensity of the electric field.

4.3.3 Mathematical Models

Due to the similar working principle and characteristic between ER damper and MR damper, the mathematical models of MR damper introduced in Section 4.2.3 can also be used to model the ER damper. In order to avoid repetition, only the viscoelastic–plastic model of ER damper is introduced in this section.

The viscoelastic–plastic models of ER dampers can be dated back to the research work of Gamota and Filisko [137], in which the response of ER materials had three distinct rheological regions: preyield, yield, and postyield. The characteristic of ER fluid is viscoelastic in the preyield region, viscoelastic–plastic in the transition through yield, and plastic in the postyield region.

4.3.3.1 Preyield mechanisms

One of the mechanical analogies that can be used to represent the viscoelastic behavior of the damper in the preyield region is Kelvin chain element [138], which is shown in Fig. 4.31A. The differential equation representing the mechanism in the time domain is

$$f_{ve}(t) = k_1 x(t) + c_1 \dot{x}(t) \tag{4.26}$$

where f_{ve} is the viscoelastic component of the damper force. Thus, the force component due to the preyield mechanism is given by

$$f_{pre}(t) = S_{ve}(\dot{x}(t)) f_{ve} \tag{4.27}$$

where S_{ve} is the nonlinear shape function or the preyield switching function that affects the smooth transition from the preyield phase to the postyield phase. The function S_{ve} is dependent on the yield velocity $\dot{x}_y$ that is chosen during the estimation process and is given by

$$S_{ve} = \frac{1}{2}\left[1 - \tanh\left(\frac{|\dot{x}| - \dot{x}_y}{4\epsilon_y}\right)\right] \tag{4.28}$$

where ϵ_y is a smoothening factor.

4.3.3.2 Postyield mechanisms

One of the mechanical analogies that can be used to represent the viscous behavior of the damper in the postyield region is the viscous damper [138], which is shown in Fig. 4.31B. The postyield force component is given by

$$f_{vi} = c_{vi} \dot{x}(t) \tag{4.29}$$

where c_{vi} is the viscous damping coefficient.

Thus, the force component due to the postyield mechanism is given by

$$f_{post}(t) = S_{vi}[\dot{x}(t)] f_{vi} \tag{4.30}$$

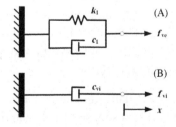

Figure 4.31
Analogs of the preyield and postyield mechanisms. (A) The Kelvin chain element and (B) the viscous damper.

where S_{vi} is the nonlinear shape function, which acts as a switching function to turn on the postyield viscous mechanism when the damper force exceeds the yield force. It is given by

$$S_{vi} = \frac{1}{2}\left[1 + \tanh\left(\frac{|\dot{x}| - \dot{x}_y}{4\epsilon_y}\right)\right] \quad (4.31)$$

4.3.3.3 Yield force

The yield force is a function of the applied field that provides the damper with semiactive capabilities. The Coulomb force or yield force effect, as seen in the damper behavior at low velocity, is described using the yield force parameter F_y

$$f_y(t) = S_y(\dot{x})F_y \quad (4.32)$$

where the shape function S_y is given by

$$S_y(\dot{x}) = \tanh\left(\frac{\dot{x}}{4\epsilon_c}\right) \quad (4.33)$$

where ϵ_c is the smoothening factor that ensures a smooth transition from the negative to positive velocities and vice versa.

In general, the viscoelastic–plastic model can be obtained by paralleling the preyield analog and the postyield analog, the damper force is given by

$$F(t) = f_{pre}(t) + f_{post}(t) + f_y(t) = S_{ve}(\dot{x})f_{ve} + S_{vi}(\dot{x})f_{vi} + S_y(\dot{x})F_y \quad (4.34)$$

where f_{pre}, f_{post}, and f_y represent the pre-, post-, and yield forces, respectively, and S_{ve}, S_{vi}, and S_y represent the shape functions given by Eqs. (4.28), (4.31), and (4.33), respectively.

4.3.4 Analysis and Design Methods

When ER dampers are installed in the structures, the equation of motion of the controlled structure is the same as Eq. (2.1). The analysis method is also the same as the structure incorporated with MR dampers. In order to avoid repetition, it is not described in this section.

4.3.5 Tests and Engineering Applications

The schematic configuration of a flow-mode ER damper [139], as shown in Fig. 4.32, consists of hydraulic and pneumatic reservoirs separated by a floating piston. Inside of the hydraulic reservoir, the piston rod is attached to a piston head. During the piston motion, ER fluid flows through a gap between inner and outer cylinder electrodes in the piston head and can be energized by applying electric fields.

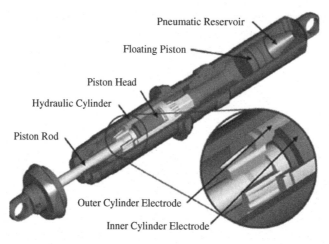

Figure 4.32
The schematic configuration of a flow-mode ER damper.

The principal design parameters are chosen as follows: electrode length is 43mm and electrode gap is 0.65 mm. On the other hand, the composition ratio of the ER fluid employed in this study is 40% corn starch and 60% peanut oil by weight. Nitrogen-filled pneumatic reservoir and floating piston located inside of the hydraulic cylinder are used to prevent cavitation on the low-pressure side of the piston and accumulate the ER fluid induced by the motion of the piston. In order to measure the field-dependent damping force of the ER damper, a hydraulic excitation system has been used [98,99,132]. The sinusoidal excitation was applied with the magnitude of 12.7 mm. The force−displacement and force−velocity hysteretic curves are similar with the results of MR dampers, as seen in Figs. 4.19−4.21 in Section 4.2.5. Li and Yang [140] designed and manufactured a type of sliding multielectrode ER fluid damper, performance test of the ER damper was carried out under sinusoidal excitation, and results showed that the equivalent damping and the energy dissipation capacity of the damper increased with increasing external electric field, and the damping characteristics of the damper were the combination of column damping and viscous damping. Liu et al. [141] carried out the performance test of ER damper under different electric field strengths and varying displacement amplitudes, and the hysteresis loops of restoring force versus displacement and velocity were experimentally obtained under a series of exciting frequencies, then a hysteretic biviscous model for the damper was established in which the parameters were determined using the testing data.

In recent years, many kinds of ER dampers of LORD Company have been launched to the market. Road test of semiactive ER-controlled vehicle suspension system fabricated by Lubrizol Company was conducted in Ford Company, which obtained good effect. In Japan, Onoda et al. used ER dampers to mitigate the vibration of a space truss structure. Park and Choi [142] studied the vibration control of a cantilevered beam via hybridization of electro-rhelogical fluids and piezoelectric films, the vibration control performances in view of the suppression of tip deflections under forced vibration were evaluated. Choi et al. [143] studied the vibration characteristics of a composite beam containing an ER fluid, the complex moduli of the composite beam were obtained by analyzing the beam's motion on free oscillation. Qu et al. [144] carried out the test and analysis of semiactive control of ER intelligent damper for structures, which showed that the dynamic response of the semiactive controlled structure was reduced more effectively than the structure incorporated with commonly used passive brace structure and the chosen of appropriate semiactive control strategy directly influenced the control effect. Liu et al. [145] developed several types of ER fluid, which have the characteristics of low viscosity at zero intensity electric field and high-viscosity gradient at varying intensity of electric field, then the tests of single mode vibration control of a sandwich plate with the developed ER fluid as the interlayer were performed, and results showed that the mechanism of vibration suppression lay in the response amplitudes, especially the resonance peaks lowered by damping effects of ER fluid. Qu and Xu [146] studied the analysis method of reticulated shells incorporated with ER dampers, the vibration equation of the controlled structure was deduced and the part semiactive control strategy was proposed, computing results showed that the ER smart damper was a kind of control device, which remains always stable and the reasonable arrangement of smart dampers on structure and the correct selection of parameters of smart dampers are very important for improving control performance. Hong et al. [147] designed and manufactured a squeeze-mode electro-rheological mount, and then the vibration control of a frame structure subjected to external excitations was studied, an optimal controller, which consists of the velocity feedback signal of the frame structure and the feedback signal transmitted from the exciting point to the mount position, was formulated in order to attenuate the imposed excitations. In addition, Hong et al. [148] studied the vibration isolation of flexible structures using this ER mounts, the mounts were incorporated with a flexible beam structure, the dynamic responses of the controlled structure were experimentally studied, and results showed that the ER mount could effectively reduce the dynamic responses.

4.4 Piezoelectricity Friction Dampers

Piezoelectric material can be manufactured into both a sensor or an actuator due to its positive and inverse piezoelectric effect. The positive piezoelectric effect refers to that when the mechanical force or deformation is applied on the material, the bound charge will appear at the two surfaces of the material, and that the charge is proportional to the applied force. Similarly, applying an electric field on the piezoelectric material will cause the production of mechanical force or deformation, which is also linear to the magnitude of the electric field. The piezoelectric friction damper mentioned in this chapter is designed based on the reverse piezoelectric effect by combining the piezoelectric actuator and the traditional friction damper, which can produce controllable friction damping force and provide a new method for the structural vibration control.

4.4.1 Basic Principles

The inverse piezoelectric effect of the piezoelectric material is used to design the piezoelectric actuator, and it will produce drive impact on the structure through its own deformation by applying the electric field. Compared with other traditional materials, the piezoelectric actuator owns the following advantages [149,150].

1. the piezoelectric material can be used as not only the sensing element but also the driving element due to its positive and reverse piezoelectric effect;
2. wide frequency response range;
3. its input and output are both electric signals, which are easy to be measured and controlled;
4. the piezoelectric effect has a good linear effect and do not need transmission mechanism, the displacement control accuracy is as high as 0.01 μm;
5. short response time with about 10 μs, which can realize the voltage follower displacement control;
6. high control force output and low-energy dissipation.

Now, the usually used piezoelectric actuators for the structural vibration control are single-layer and multilayer types. The former is usually pasted on the surface of the flexible beam, plate, and shell for vibration control due to its small driving force. The latter can effectively solve the problem of small driving force and displacement by composition of many piezoelectric actuators.

If the piezoelectric actuator is combined with the passive friction damper, the electric-induced deformation of the piezoelectric ceramic can actively adjust the fastening force between the friction elements according to the requirement of structural vibration control and realize the real-time adjustment of the friction force, which attributes the intelligent characteristic to the friction damper.

4.4.2 Construction and Design

Based on the working mechanism, a new piezoelectric friction damper [151] is designed, by Xu et al. as shown in Fig. 4.33, this damper is composed of two parts, including passive and active energy consumption. The passive part consists of preloading spring, piston, and working cylinder, and the preloading force can be adjusted according to the actual condition. The material used between the piston and working cylinder is ceramic friction plates, of which the friction coefficient is very high. The active part consists of piezoelectric rod and power supply system, the piezoelectric rod is connected to the piston through connected element, the cross-sectional area of the piezoelectric rod can be determined according to the required maximum damping force.

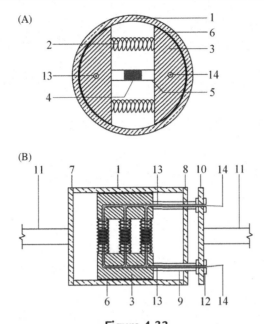

Figure 4.33
Diagram of the piezoelectric friction damper. (A) Sectional view and (B) profiles.

As shown in Fig. 4.33, the name of each part is: (1) working cylinder, (2) preloading spring, (3) piston, (4) piezoelectric organic matter, (5) connected rod, (6) high-damping friction material, (7) and (8) limit baffle plate, (9) connected bar, (10) connected steel plate, (11) connected element, (12) steel plate, (13) hole, (14) wire.

4.4.3 Mathematical Models

Due to the normal stress applied on the cylinder is far less than the elastic limit of the outer cylinder and ceramic, the outer cylinder and piston are assumed both absolutely rigid. It can be seen that the radial stress applied on the outer cylinder wall is uneven as well as the stress applied on the piston surface due to the outer cylinder wall, and the stress is large in the middle and small at the two ends.

Fig. 4.34 shows the computing model of the piston, in order to calculate the damping force of the piezoelectric friction damper, the radial distribution force of the piston string surface should be determined first. Assuming that the radial distribution force $q_{(\theta)} = q_0 f(\theta)$, then the following equations can be obtained according to $\sum N_y = 0$,

$$\int_{-\alpha}^{\alpha} q_0 \cos \theta \, R d\theta \cdot \cos \theta = 2F \tag{4.35}$$

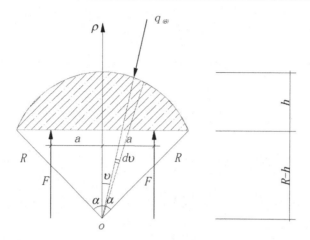

Figure 4.34
Computing model of the piston.

Table 4.3: Parameters of the friction damper

Radius, R (m)	String Height, h (m)	Concentration Force, F (N)	Distance Between Concentration Force and Circle Center, a (m)
0.3	0.2	5000	0.05

$$\alpha = \arccos \frac{R-h}{R} \tag{4.36}$$

$$q_0 = \frac{4F}{R(\sin 2\alpha + 2\alpha)} = \frac{2F}{((R-h)/R)\sqrt{(2R-h)h} + R\arccos((R-h)/R)} \tag{4.37}$$

$$N(t) = \int_{-\alpha}^{\alpha} q_0 \cos\theta \, R\mathrm{d}\theta = 2q_0 R \sin\alpha = \frac{8F \sin\alpha}{(\sin 2\alpha + 2\alpha)}$$

$$= \frac{4F\sqrt{(2R-h)h}}{((R-h)\sqrt{(2R-h)h}/R) + R\arccos((R-h)/R)} = k \times F \tag{4.38}$$

$$k = \frac{4\sqrt{(2R-h)h}}{\frac{(R-h)\sqrt{(2R-h)h}}{R} + R\arccos\frac{R-h}{R}} \tag{4.39}$$

In order to verify the accuracy of the assumed radial distribution force $q_{(\theta)}$, the finite element simulation is carried out, the basic parameters are listed in Table 4.3. The piston string surface is modeled using 72 elements, and the third principle $q_{(\theta)}$ stress is chosen as the radial pressure of each point, as shown in Fig. 4.35, and then the resultant radial

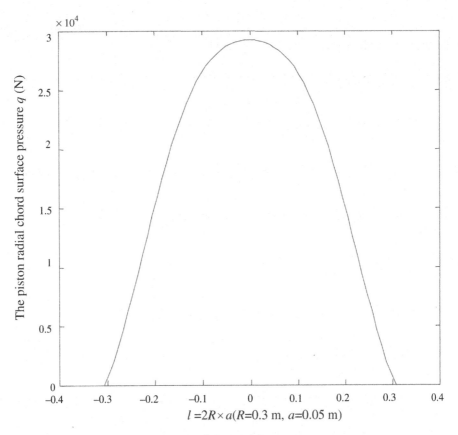

Figure 4.35
Distribution of the radial force of the piston string surface.

pressure force can be obtained to 11,465 N, Theoretical resultant force can be also obtained up to 12,203 N by using Eqs. (4.38) and (4.39). The relative error is 5%, which shows that the assumed radial distribution force $q_{(\theta)}$ is reasonable. Then, Eqs. (4.38) and (4.39) can be used to compute the final force of the piezoelectric friction damper in the following.

Assuming F_P is the constraint force of the piezoelectric actuator under the electric field E, then,

$$\begin{cases} F_P = \varepsilon_P Y_P A_p \\ \varepsilon_p = \dfrac{\Delta L_E}{L_p} \end{cases} \quad (4.40)$$

where $\Delta L_E = d_{33} E L_p$ is the elongation of the actuator under the electric field E, d_{33} is the axial piezoelectric strain constant, Y_P is the elastic modulus of the piezoelectric ceramic, A_p is the cross-sectional area of the piezoelectric ceramic actuator, L_p is the axial length of the piezoelectric ceramic actuator.

$$F = F_P + F_S \tag{4.41}$$

where F_S is the preloading force provided by spring.

Then, the positive pressure applied on the piezoelectric friction damper can be written as

$$N(t) = k \times d_{33} E Y_p A_p + k \times F_s \tag{4.42}$$

In order to simplify this expression, there is,

$$N(t) = K d_{33} E + N_0 \tag{4.43}$$

where $K = k \times Y_p \times A_p$ is the shape coefficient of the piezoelectric friction damper, $N_0 = k \times F_s$ is the pressure provided by spring. Then, the positive pressure applied on the piezoelectric friction damper can be expressed as follows

$$N(t) = \begin{cases} 2nN_0 & (E = 0) \\ 2nKd_{33}E + 2nN_0 & (E > 0) \end{cases} \tag{4.44}$$

where n is the number of spring and piezoelectric actuator.

Thus, the damping force model of the piezoelectric friction damper can be written as

$$f(t) = \mu N(t) \, \text{sgn}[\dot{x}(t)] \tag{4.45}$$

where μ is the friction coefficient between the piston and the inner surface of the outer cylinder, $\dot{x}(t)$ is the relative velocity of the piezoelectric friction damper.

The basic parameters of the damper are listed in Table 4.4. The friction coefficient of the damper is $\mu = 0.5$, preloading force of the spring is 5000 N, the number of piezoelectric actuator is 2, maximum voltage is 300 V, and maximum electric field intensity $E_0 = 7.5 \times 10^5$ V/m. Based on the above analysis, the maximum damping force of the piezoelectric damper can be calculated up to 10.13 t. The theoretically computed hysteresis curves are depicted, as shown in Fig. 4.36. It can be seen that the working voltage can be adjusted real-time to get the damping force for structural vibration control.

Table 4.4: Basic parameters of the piezoelectric friction damper

Piston		Piezoelectric Actuator			
Radius (m)	Height (m)	Height (m)	Radius (m)	Elastic Modulus (Pa)	Axial Piezoelectric Strain Coefficient (m/V)
0.30	0.20	0.1	0.025	6×10^{10}	1×10^{-10}

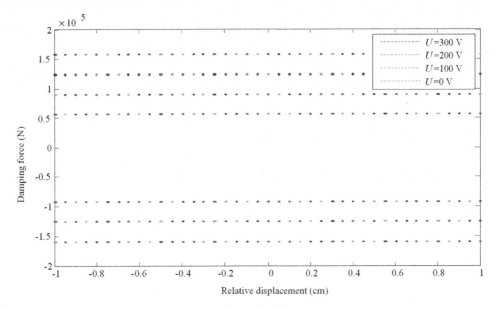

Figure 4.36
The hysteresis curves of the piezoelectric friction damper.

4.4.4 Analysis and Design Methods

Assuming that a total of p piezoelectric friction dampers are installed in a structure with n degrees of freedom, the equation of motion of the controlled structure can be expressed as Eq. (2.1). The control force of the ith damper is in the following type:

$$\{f_d(t)\} = \{f_{dy}\}\text{sgn}(\dot{x}_d) \tag{4.46}$$

where $\dot{x}_d$ is the relative velocity of the piezoelectric damper, $\{f_{dy}\}$ is the variable damping force of the piezoelectric damper, and it depends on the property of the actuator and input voltage as well as the preloading force of the friction damper and the friction coefficient. The corresponding control algorithm has been introduced in the previous chapter.

A three-story reinforced concrete structure is chosen to analyze the vibration control effect of the piezoelectric friction damper, one piezoelectric friction damper is supposed to added in each floor of a three-story reinforced concrete structure, as shown in Fig. 4.37. The masses of the first, second, and third floor are $m = 1.690 \times 10^4$ kg, 1.581×10^4 kg, 1.412×10^4 kg, respectively. The stiffness of the first, second, and third floor are $k = 1.512 \times 10^7$ N/m, 2.012×10^7 N/m, 2.012×10^7 N/m, respectively. The Rayleigh damping ratio is adopted in this analysis procedure, the El Centro earthquake wave with a magnitude of 200 gal and the time step of 0.02 s is chosen as the excitation input. The

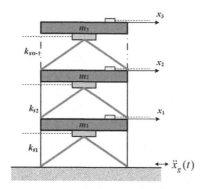

Figure 4.37
The diagram of the controlled structure.

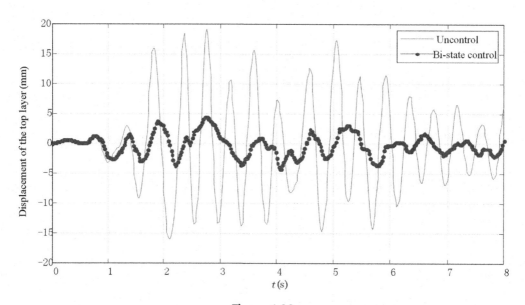

Figure 4.38
The displacement response of the top layer of the structure.

maximum damping force provided by the damper is 158.64 kN. The analysis results are shown in Figs. 4.38 and 4.39, from which we can see that both the acceleration responses and the displacement responses are reduced significantly.

4.4.5 Tests and Engineering Applications

In recent years, piezoelectric material has been widely used in the aerospace and mechanical structure. However, piezoelectric material is mainly used as sensors in the

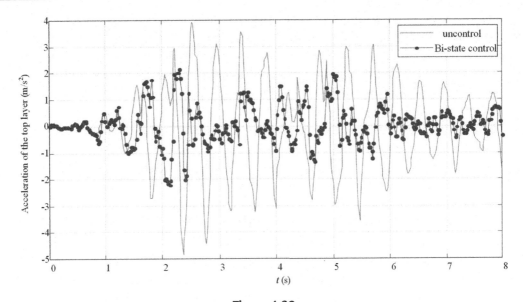

Figure 4.39
The acceleration response of the top layer of the structure.

bridge and building health monitoring in civil engineering [152–156]. Some researchers used the material to design the piezoelectric friction dampers to control the dynamic responses of the engineering structures. Qu [157] designed a piezoelectric friction damper by combining the Pall friction damper with the piezoelectric ceramic washer, of which the fastening force of the bolt can be adjusted by changing the electric-induced deformation of the washer to realize the changing of damping force, then Ou et al. [158,159] developed a T-shape PZT variable friction damper and conducted the performance test. Qu et al. [160] and Xu et al. [161] designed the friction bar with adjustable parameters by combining the piezoelectric material with the custom friction damper. Under the electric field, the deformation of the piezoelectric material is constrained, which leads to the adjustable normal pressure, thus the friction force can be adjusted real-time according to the dynamic response of the structure, then this damper was used in the practical engineering project to suppress the wind-induced vibration of Hefei Jade TV tower. Garrett et al. [162] did a lot of static and dynamic experiments on the piezoelectric friction damper, and obtained the force–displacement hysteresis curves, and the friction coefficient, equivalent damping ratio, and effective stiffness are obtained according to the experimental data, the fatigue test was also performed, the results show that the piezoelectric friction damper has a very good fatigue resistance properties. Chen et al. [163,164] also conducted the performance test of piezoelectric friction damper, and carried out the shaking table test of a quarter-scale three-story steel frame structure incorporated with piezoelectric friction dampers, results show that the piezoelectric friction damper can effectively reduce the dynamic responses of the structure. Li et al. [165,166] designed a new type piezoelectric friction damper by combining the stack type tubular piezoelectric ceramics driver and the Slotted Bolted Connection (SBC) friction damper proposed by Song [167], the prepressure was applied on the actuator through fastening bolt to restrict the deformation of the actuator, then the piezoelectric actuator can provide varied control force to realize semiactive control. Dai et al. [168] designed a piezoelectric varied friction damper by considering the commercially available multilayer piezoelectric stack actuators and the circular friction disc, this damper can provide varied damping force in each horizontal direction, and it can work together with circular seismic isolator and form smart isolation system. This damper was fabricated and performance test on this damper was also carried out. Later, the piezoelectric–Shape Memory Alloy (SMA) complex varied friction dampers [169] was designed, and experimental validation was also carried out by using a series of shaking table tests on a four-story steel isolated structure model, results showed that the damper was effective in mitigating earthquake responses on both passive-off control and the semiactive control. Wang et al. [170,171] designed a piezoelectric friction damper and conducted the performance test under different prepressure, then earthquake shaking table test was carried out to verify the control effect of the piezoelectric dampers on the dynamic responses of a transmission tower structure under El-Centrol wave excitation, results showed that the piezoelectric dampers could effectively reduce the dynamic responses of the structure.

4.5 Semiactive Varied Stiffness Damper

4.5.1 Basic Principles

The working principle of semiactive variable stiffness vibration control technology is on the basis of changing the additional stiffness of the variable stiffness device in the controlled structure, so as to avoid the natural frequency of the controlled structure system being too close to or same as the excitation frequency, then to achieve the purpose of avoiding resonance of the whole system, finally to ensure that the structure is in the states of stability and security. From the point of view of energy conversion, the semiactive varied stiffness control is achieved through the deformation of the variable stiffness component so that the part of the structure vibration energy can be converted to the elastic deformation energy of the stiffness components. Then, the absorption elastic deformation energy of these stiffness elements can be released (actually, the elastic deformation energy is converted to the heat energy of the servo system), and at the same time, the part of the vibrational energy can be consumed by the damping components. Therefore, the vibration of the structure is reduced.

A typical semi-active variable stiffness control device includes three parts: additional stiffness element, mechanical device, and controller, as shown in Fig. 4.40 [172].

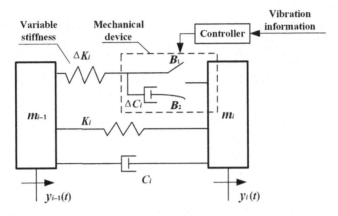

Figure 4.40
Configuration of the variable stiffness system.

According to the additional stiffness component is connected or not, the variable stiffness system can produce two kinds of states: ON and OFF. The changes of the additional stiffness states (ON or OFF) can be realized by switching the controller states of the variable stiffness device according to the control algorithm.

4.5.2 Construction and Design

Based on the working principle of the semiactive variable stiffness control technology, some variable stiffness devices, as shown in Fig. 4.41, have been designed and experimentally investigated by many researchers [99,173,174]. These semiactive stiffness control devices can provide adjustable stiffness, thus the natural dynamic characteristics of the controlled structure attached to these semiactive devices can be changed.

The control force (F_D) of the variable stiffness device can be got from Eq. (4.47), which is calculated by analyzing the whole control system.

$$\Delta K_{eff} x_d(t) = F_D \to F_{opt} \qquad (4.47)$$

where ΔK_{eff} is the equivalent additional stiffness provided by the variable stiffness device, $x_d(t)$ is the displacement of the structure, and F_{opt} is the optimal control force of the system.

For designing the variable stiffness device, the maximum output F_D is the most important index, which affects vibration control result of the structure. Take the electro-hydraulic variable stiffness device for example [175], the output force F_D is associated with the piston diameter, hydro-cylinder diameter, and the oil characteristics. According to the calculated F_{opt}, the parameters of the variable stiffness device can be optimized to get the suitable output F_D, which is close to the optimal control force F_{opt}.

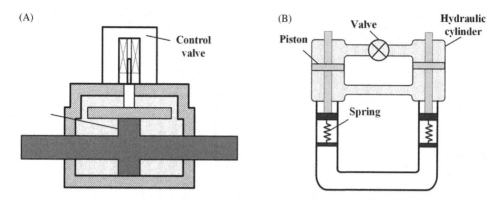

Figure 4.41
Variable stiffness device. (A) Semi-active variable stiffness damper and (B) tuned liquid column damper.

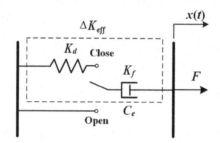

Figure 4.42
Model of the semiactive variable stiffness system.

4.5.3 Mathematical Models

As the name suggests, the semiactive variable stiffness damper could provide additional stiffness to the controlled structure, that is, the semiactive variable stiffness damper is able to switch dynamically the working condition of the control valve, thereby dissipation or absorption of energy in the structure will take place. The parameter model of the electro-hydraulic variable stiffness damper mechanism is shown in Fig. 4.42.

These dampers can produce a large control action by dynamically changing its stiffness characteristics. When the control valve is closed, the damper serves as a stiffness element and will provide an additional stiffness to the structure. The effective stiffness ΔK_{eff} consists of K_f (provided by the bulk modulus of the fluid in the hydraulic cylinder) and K_d (the stiffness provided by the springs shown in Fig. 4.41B and the bracing connected with the damper). When the control valve is open, the piston is free to move and the damper will provide only certain damping (C_e) without stiffness. The stiffness ΔK_{eff} is the equivalent effective stiffness of the entire damper and can be calculated by

$$\Delta K_{eff} = \frac{K_d K_f}{K_d + K_f} \tag{4.48}$$

4.5.4 Analysis and Design Methods

The schematic model of the semiactive variable stiffness damper used for structural vibration control of the single-freedom system is shown in Fig. 4.43, and the equation of motion of the controlled structure can be expressed as follows

$$m\ddot{x}(t) + c_s\dot{x}(t) + (k_s + \Delta k^*)x(t) = -m\ddot{x}_g(t) \tag{4.49}$$

where k_s, c_s separately are the stiffness and damping of the structure; $\ddot{x}_g(t)$ is the input excitation; and Δk^* is the additional stiffness provided by the semiactive variable stiffness damper.

The variable elastic force $\Delta k^* x(t)$ can be moved to the right side of the motion equation as the semiactive control force, $f_d = \Delta k^* x(t)$ in Eq. (4.50), which is similar to the active control force in the active variable stiffness control system.

$$m\ddot{x}(t) + c_s\dot{x}(t) + k_s x(t) = -m\ddot{x}_g(t) - f_d(t) \tag{4.50}$$

According to the principle that the control forces are equal, $f_d = f_{max}$, in which f_{max} is the optimal control force, the stiffness of the variable stiffness device can be set as

$$\Delta k^* = \frac{f_{max}}{x_{max}} \tag{4.51}$$

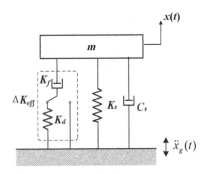

Figure 4.43
Analysis model of the semiactive variable stiffness system.

And the additional stiffness Δk^* can be written as follows by using the switching control law

$$\Delta k^* = \begin{cases} \Delta k_{\text{eff}} & \text{OFF} \\ 0 & \text{ON} \end{cases} \tag{4.52}$$

where $\Delta k^* = \Delta k_{\text{eff}}$, if the switch is closed; and $\Delta k^* = 0$, if the switch is open. The equivalent effective stiffness Δk_{eff} can be calculated from Eq. (4.48).

There are p semiactive variable stiffness devices installed in an n stories structure, as shown in Fig. 4.44. The equations of motion of the linear, n-story building structure equipped with semiactive variable stiffness system can be expressed as follows

$$\begin{cases} [M]\{\ddot{x}(t)\} + [C]\{\dot{x}(t)\} + [K]\{x(t)\} = \{P(t)\} + [B]\{f_d(t)\} \\ x(t_0) = x_0 \quad \dot{x}(t_0) = \dot{x}_0 \end{cases} \tag{4.53}$$

where the meaning of each item is the same as Eq. (2.1).

Eq. (4.53) can be represented in the state-space form as

$$\begin{cases} \{\dot{Z}(t)\} = [A]\{Z(t)\} + [D_0]\{P(t)\} + [B_0]\{f_d(t)\} \\ \{Y\} = [E_0]\{Z(t)\} \end{cases} \tag{4.54}$$

where the meaning of each part is the same as Eq. (2.5).

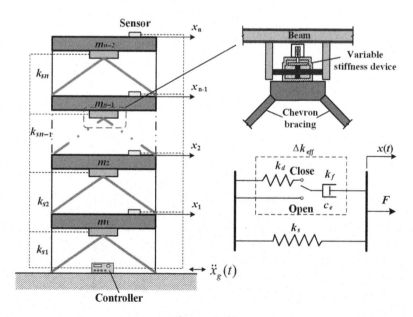

Figure 4.44
Analysis model of the semiactive variable stiffness system.

The semiactive control force of these semiactive variable stiffness dampers u_i ($i = 1, 2, \ldots, p$) can be calculated as

$$u_i(t) = \Delta k_i g_i(t) x_i(t) \tag{4.55}$$

In Eq. (4.55), the $g_i(t)$ expressed for switching control law as

$$g_i(t) = \begin{cases} 1 & (\dot{x}_i x_i > 0) \\ 0 & (\dot{x}_i x_i \leq 0) \end{cases} \tag{4.56}$$

Therefore, the variable stiffness Δk_i of the dampers can be calculated according to the equivalent principle of which the control force is equal, as seen Eq. (4.51).

4.5.5 Tests and Engineering Applications

Generally, a semiactive stiffness damper consists of a fluid-filled cylinder, a piston, and a motor-controlled valve. Based on the feedback information of structural responses, the motor regulates the valve open or close to dissipate or absorb vibration energy of the structures. A significant amount of researches and developments have been conducted on semiactive variable stiffness devices. Li and Liu [176,177] conducted a shaking table test of semiactive structural control using variable stiffness, which is set in the 1:4 model of a five-story frame, the test results showed that this type of control system was capable of efficiently reducing the structural vibration response. Kobori and Nasu [178,179] also conducted shaking table test of the structure incorporated with the variable stiffness system, results showed that the device used was efficient. Kumar et al. [180] used the semiactive control variable stiffness damper (SAVSD) to control the vibration of piping system used in industry, fossil and fissile fuel power plant, and the SAVSD changed its stiffness depending upon the piping response and accordingly added the forces in the piping system. Nasu et al. [181] experimentally studied a high-rise building with variable stiffness system. The model of the building is a 25-floor steel frame structure, and test results showed that the variable stiffness system needed only a little external energy and can reduce the responses of the structure efficiently. The resetting semiactive stiffness dampers designed for controlling a 2D, three-story, three-bay frame under random excitations were experimentally studied by Jabbari and Bobrow [182]. A semiactive control system with the switching semiactive stiffness dampers used for controlling a 2D, 20-story frame under two different earthquake waves were studied by Kurino et al. [183]. Similarly, some other tests on the semiactive control systems with different semiactive stiffness dampers and different control strategies have been conducted in succession [184–186].

The varied stiffness system [187] was designed to prevent resonance with the seismic motion, by means of controlling the stiffness of the building through a control computer (nonstationary, nonresonant), according to the ever-changing properties of the seismic

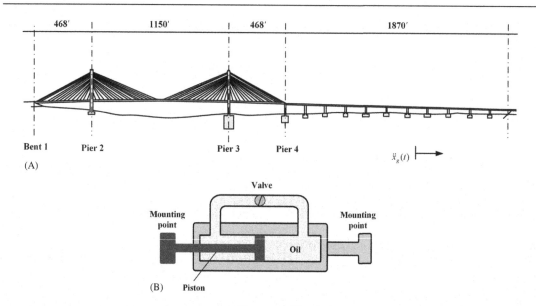

Figure 4.45
Schematic diagram of the bridge and semiactive variable stiffness damper. (A) Schematic drawing of the cable-stayed bridge and (B) schematic diagram of the semi-active variable stiffness damper.

motion. The building was thus protected from the destructive energy of the earthquake. The control computer instructed the varied stiffness damper how to alter the stiffness of each story to reduce the response of the building by preventing its resonance with the seismic motion, through the selection of different stiffness types. Application of the semiactive variable stiffness damper has been investigated for a cable-stayed bridge, shown in Fig. 4.45A [188]. In the research, a resetting semiactive stiffness damper, shown in Fig. 4.45B, was used to control the peak dynamic response of the cable-stayed bridge subject to earthquakes. It is shown that the designed damper is quite effective on reducing peak response quantities of the bridge. The tests and application of previous studies indicate that the semiactive variable stiffness device is effective in reducing structural responses.

4.6 Semiactive Varied Damping Damper

As a classical example of semiactive control system, semiactive varied damping dampers can adjust the structural damping in real-time by means of the additional damping device, which is installed at the deformation members of the structures. Generally, semiactive varied damping dampers consist of traditional fluid damper or viscous damper and a controllable servo valve, and the damping force can be adjusted by controlling the servo valve to changing the fluid flow. This method is very effective and can reduce the structural vibration significantly. In this section, the varied damping damper will be discussed in detail.

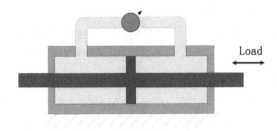

Figure 4.46
Sketch of the oil damper with adjustable aperture.

4.6.1 Basic Principles

The varied damping damper system is firstly proposed by Hrovat [98], the damping is adjusted in real-time by using the additional damping device. The classical varied damping damper is shown in Fig. 4.46. It consists of a hydraulic cylinder, a piston, and an electro-hydraulic servo valve. Only very small energy is needed to adjust the opening size of the servo valve, and the damping force can be achieved to 100–200 t.

The damping force provided by the varied damping damper depends on the opening size of the servo valve. As shown in Fig. 4.46, when the valve is completely open, the damper can provide minimum damping force, named the passive-off state. On the contrary, the maximum damping force is provided when the servo valve is completely closed, which is called the passive-on state. The range of damping force provided by the damper is between the passive-off and passive-on state, and can be adjusted according to the dynamic response of the structure.

4.6.2 Construction and Design

Based on the working principle of the semiactive variable damping control technology, some variable damping devices have been designed and experimentally investigated by many researchers [189–192]. These semiactive variable damping dampers are utilized to adjust the damping so that the dynamic responses of the structure can be reduced to meet the demand.

Symans and Constantinou [193] designed a variable damping damper based on the passive viscous fluid damper, as shown in Fig. 4.47. This device consists of oil cylinder with oil inlet and outlet, piston with small holes, bypass pipeline with the servo control system and the accumulator. The servo control system is a coil using closed-loop control. When the small hole of the bypass pipe is closed, the fluid will flow directly from one oil chamber to the other oil chamber through the small hole in the piston, and will not flow from the bypass pipe. At this time, this device is equivalent to the traditional passive viscous fluid

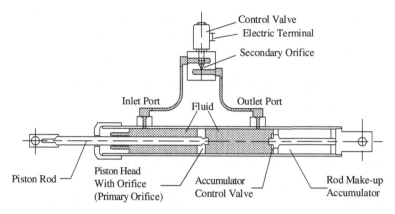

Figure 4.47
Construction of the varied damping damper.

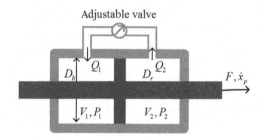

Figure 4.48
Computing model of the varied damping device.

damper and can provide the largest damping coefficient $c_{d\ max}$. When the small hole in the bypass pipe is opened completely, then the fluid will flow from one chamber to another chamber through both holes in the piston and the bypass pipe. The device is also equivalent to the traditional passive viscous fluid damper, but it provides the smallest damping coefficient $c_{d\ min}$. By adjusting the opening size of the hole of the bypass pipe, the damping coefficient of the device varies from the smallest with closed holes to the largest with completely open holes in the bypass pipeline.

4.6.3 Mathematical Model

The variable damping damper is actually like a hydraulic system, thus the mathematical model can be established using the fluid mechanics theory according to the oil circuit structure. The classical variable damping device will be employed to introduce the modeling method [194], and the computing model is shown in Fig. 4.48.

Assuming the force applied on the piston is u_d and the relative velocity is $\dot{x}$, the effective area of the piston is A_p, the pressure of the two cavities are p_1, p_2, and the corresponding volumes are V_1, V_2. The damping coefficient is c_v (adjustable, with respect to voltage).

According to the force balance of the piston rod, the control force of the device can be written as follows:

$$F = A_p(p_2 - p_1) \tag{4.57}$$

Assuming the fluid in the servo valve is incompressible, the pressure is linear to the flow, then,

$$Q_1 = Q_2, \quad p_2 - p_1 = c_v Q \tag{4.58}$$

Assuming the fluid in the hydraulic cylinder is compressible, the volume elastic modulus is β, then the volume change due to the pressure change is

$$\dot{p}_1 V_1 = -\beta \dot{V}_1, \quad \dot{p}_2 V_2 = -\beta \dot{V}_2 \tag{4.59}$$

where the minus sign represents the decrease of fluid volume.

The flow change rates are as follows:

$$\dot{V}_1 = -Q_1 + A_p \dot{x}, \quad \dot{V}_2 = Q_2 - A_p \dot{x} \tag{4.60}$$

Based on the above equations, the pressure change rate in the two cavities due to the motion of the piston rod is [194]

$$\dot{p}_2 = -\frac{\beta}{V_2} \frac{F}{A_p c_v} + \frac{\beta}{V_2} A_p \dot{x} \tag{4.61}$$

$$-\dot{p}_1 = -\frac{\beta}{V_1} \frac{F}{A_p c_v} + \frac{\beta}{V_1} A_p \dot{x} \tag{4.62}$$

Combining the above two equations, the rate of the pressure drop is

$$\dot{p}_2 - \dot{p}_1 = -\frac{\beta}{A_p c_v}\left(\frac{1}{V_1} + \frac{1}{V_2}\right) F + \beta A_p \left(\frac{1}{V_1} + \frac{1}{V_2}\right) \dot{x} \tag{4.63}$$

According to Eq. (4.57), the controllable control force of the varied damping damper can be written as follows:

$$\dot{F} = \frac{k_d}{c_d} F + k_d \dot{x} \tag{4.64}$$

where $k_d = 4\beta A_p^2/V_T$ is the stiffness of the hydraulic system, $V_T = 4((1/V_1)+(1/V_2))^{-1}$, $c_d = c_v A_p^2$ is the damping coefficient of the hydraulic system. Eq. (4.64) is the final computing equation of the damping force of the varied damping device.

4.6.4 Analysis and Design Methods

The schematic model of the semiactive variable damping damper used for structural vibration control of the single-freedom is shown in Fig. 4.49. According to the damping force type provided by the device, the motion equations of the controlled structure can be expressed as follows

$$\text{uncontrolled } m\ddot{x} + c\dot{x} + kx = f \quad (4.65)$$

$$\text{Passive-on } m\ddot{x} + (c + c_{d\ max})\dot{x} + kx = f \quad (4.66)$$

$$\text{Passive-off } m\ddot{x} + (c + c_{d\ min})\dot{x} + kx = f \quad (4.67)$$

$$\text{Semi-active } m\ddot{x} + c\dot{x} + kx + c_d(v,t)\dot{x} = f \quad (4.68)$$

where f is the disturbance, $c_d(v,t)$ is the damping coefficient with respect to voltage and time provided by the varied damping damper. A three-story frame structure is adopted to analyze the control effect of varied damping dampers. The mass and story stiffness are $m_i = 4 \times 10^5$ kg, $k_i = 2 \times 10^8$ N/m, respectively. The Rayleigh damping ratio is adopted and the first two damping ratios are both 0.05. El Centro wave is employed with a peak value of 200 gal. Three varied damping dampers are arranged at each story, the maximum damping coefficient of the three semiactive varied damping dampers are $c_{1d\ max} = 7.7012 \times 10^6$ N·s/m, $c_{2d\ max} = 7.7232 \times 10^6$ N·s/m, $c_{3\ dmax} = 5.42 \times 10^6$ N·s/m, the minimum damping coefficients of the three dampers are $c_{1d\ min} = 9.6265 \times 10^5$ N·s/m, $c_{2d\ min} = 9.6539 \times 10^5$ N·s/m, $c_{3d\ min} = 6.7716 \times 10^5$ N·s/m. The additional damping ratio due to the damper (passive-off state) is 2.34%, the maximum responses under different control strategy are listed in Table 4.5.

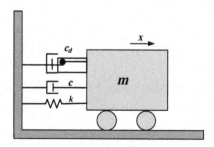

Figure 4.49
Schematic model of the controlled structure.

Table 4.5: The maximum responses and control forces of varied damping control

Control Algorithm	Maximum Story Displacement (cm)			Maximum Acceleration (m/s²)			Maximum Control Force (kN)		
	1	2	3	1	2	3	1	2	3
Without control	2.37	2.07	1.20	3.72	4.49	5.99	–	–	–
Passive-off	2.20	1.89	1.09	3.15	4.17	5.43	188.87	168.93	86.93
Semiactive	1.44	1.16	0.65	2.46	3.10	3.48	940.77	731.20	417.13

In order to realize semiactive control, the optimal Bang-Bang control strategy, which can be expressed as:

$$c_d(t) = \begin{cases} c_{d\ max} & u\dot{x} < 0 \\ c_{d\ min} & u\dot{x} \geq 0 \end{cases} \quad (4.69)$$

According to Table 4.5, results show that both the displacement and acceleration responses of the structure are reduced by using passive-off control and semiactive control. Comparatively speaking, the passive-off control method can only reduce the dynamic responses in a small extent, while the acceleration and displacement responses of the structure using semiactive control can be reduced by 30–40%, which shows that semiactive control can achieve excellent control effect. In addition, the control effect of passive-off algorithm is dominated by the design of the minimum damping coefficient $c_{id\ min}(i = 1, 2, 3)$. Apparently, the control effect will be more significant with increasing $c_{id\ min}$.

4.6.5 Tests and Engineering Applications

A significant amount of research and developments have been conducted on semiactive variable damping devices. Kawashima [192], Symans and Constantinou [193], and Niwa [190] conducted a series of mechanical tests on the varied damping dampers under different excitation frequencies and amplitudes, and obtained the force–displacement hysteresis curves. Symans [193] conducted a shaking table test on a three-story steel frame structure incorporated with variable damping dampers. The geometric similarity ratio between the model and the prototype is 1:4, the whole mass of the structure is 2868 kg with 956 kg for each layer. The variable damping device is installed at the bottom of the structure, and the dynamic responses can be measured using the acceleration sensors and displacement sensors. Research results show that the variable damping devices can significantly reduce the dynamic responses of the structures. Niwa and Kobori [190] used the variable damping damper to control a steel office building located at Shizuoka city. Eight dampers were installed at the gable of one to four layers, and results show that the designed damper can significantly increase the damping of the structure and is quite effective in reducing peak response quantities of the steel structure. As we know, the long-time vehicle loading on the bridge will lead to the fatigue effect, thus the variable damping damper is especially suitable for vibration control of the bridge structure. Kawashima [192], Shinozuka [195], Patten [196], and Gavin [194,197] conducted a lot of researches on the vibration control of bridges using variable damping device. Patten firstly took the variable damping device into actual engineering project to reduce the vibration of a I-35 Bridge in America, and results show that the semiactive variable damper has an excellent control effect, and the minimum service life can be extended by 35.8 years. Li et al. [198] developed a kind of fluid viscous semiactive damper and experimentally studied the performance of the controller of the semiactive damper, then a 1:4 scale five-story steel structure with and without the semiactive damper was tested using a shaking table under EL-Centro and Tianjin earthquake ground excitations, in which the control laws of Hrovat and ON/OFF algorithms were used. The results indicated that semiactive damper reduced earthquake responses significantly. Yang et al. [199] studied the effectiveness of using a semiactive variable damper to control seismically excited structures, and results show that the performance of variable dampers in reducing the seismic response of structures depends on the ratio of the disturbance frequencies to the natural frequencies of structures, rather than the natural frequencies of structures only.

The above tests and application of previous studies indicated that the semiactive variable damping damper was effective in reducing structural responses.

4.7 MRE Device

MRE is a kind of smart material, whose mechanical properties, such as stiffness and damping, can be changed continuously, rapidly, and reversibly by an applied magnetic field [200–204]. The physical phenomena of MR elastomers are very similar to that of MR fluids. Simply, MREs are solid-state analogs of MR fluids, where the liquid is replaced by a rubber material. MREs generally consist of three main components: matrix, magnetic particles, and additives. The advantage of MREs over MR fluids is that magnetic particles do not undergo sedimentation; therefore, these MREs have both the MR effect and good mechanical performance [205]. Such properties stimulate their many promising applications such as vibration absorbers, variable stiffness devices, and variable impedance surfaces. Compared with MR fluids, the application of MRE is still at an exploring stage.

4.7.1 Basic Principles

It can be seen from the composition of MRE that the MRE is a kind of composite material, which is made of magnetized particles dispersed in a solid polymer medium such as rubber. Under the external magnetic field, the matrix is cured, then the magnetic particles are locked into a place and the chains are firmly established in the matrix. The elastic modulus of an MRE increases monotonically with the strength of the magnetic field. Upon removal of the magnetic field, the MRE immediately reverts to its initial status. The changed modulus of MRE caused by a magnetic field is called magnetic-induced modulus, which affects the adjustability of the MRE, so it is an important index of MRE properties.

In order to explain the causes of the magnetic effect of MREs and provide the theoretical basis for improving the performance of MRE. The microphysical model is proposed by some researchers from different angles. This kind of microphysical model, such as magnetic dipole model, coupling field model, single chain model and so on, can simulate the magnetic-induced modulus of MREs.

Magnetic dipole model [206]:

$$\Delta G = \frac{(4\gamma^4 - 27\gamma^2 + 4)\Phi J_p^2}{8\mu_1\mu_0 r_0^3(\gamma^2+1)^{9/2}} \quad (4.70)$$

Coupling field model [207]:

$$\Delta G = \frac{9}{8}\frac{\Phi C|m_i|^2(4-\gamma^2)}{r_0^3\pi^2 d^3 \mu_1\mu_0(1+\gamma^2)^{7/2}} \quad (4.71)$$

Single chain model [208]:

$$\Delta G = \frac{9\Phi|m_i|^2\zeta}{4\pi^2 d^3 r_0^3 \mu_1\mu_0} \quad (4.72)$$

where ΔG is the magnetic-induced modulus of the MRE, Φ is the volume ratio of magnetic particles, r_0 is the distance between two magnetic particles, d is the magnetic particle diameter, γ is the shear strain, $|m_i|$ is the magnetic dipole moment, and J_p is the intensity of the polarization of magnetic particles.

It can be seen from the microphysical model that the magnetic modulus is affected by many factors, such as the volume ratio of magnetic particles Φ, the distance between two magnetic particles r_0, the magnetic particles diameter d, the shear strain γ, the external magnetic field H, and so on. The Φ, r_0, and d decide the performance of MRE when the MRE is prepared, whose magnetic modulus mainly affected by the shear strain γ and the external magnetic field H.

According to different stress states, the operation modes of MRE can be divided into shear mode and tension-compression mode. The schematic diagram of MRE's operation mode is shown in Fig. 4.50.

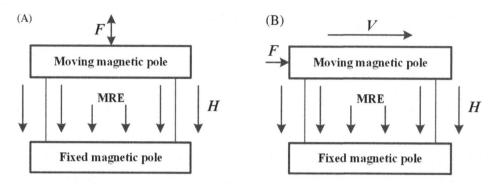

Figure 4.50
MRE's operation modes. (A) Tension-compression mode and (B) shear mode.

The tension-compression mode, as shown in Fig. 4.50A, in which the pole plate carry through relative movements in the direction parallel to the field, and the MRE is in the alternation state of tensile and compression. Due to the relative movement of the magnetic pole plates, the magnetic field strength, between the two pole plates, changes continuously, so as to achieve the purpose of changing the properties of MRE. However, due to the coupling between the magnetic nonlinear of magnetic field and magnetic particles, the MRE device system with tension-compression mode is very complicated and it is difficult to design, therefore, there is few devices designed in the tension-compression mode.

The MRE in shear mode placed in the two pole plates is shown in Fig. 4.50B: one pole plate is fixed and another moves in the direction perpendicular to magnetic field. So, the mechanical properties can be changed by controlling the external magnetic field. Due to that the distance between the two pole plates is very small and constant, then, it is easy to form a relatively uniform magnetic field, so it is relatively simple to design the MRE device system with the shear mode. At present, most of the MRE devices adopt the shear operation mode.

4.7.2 Construction and Design

Based on the shear operation mode, different kinds of MRE devices are designed and studied, such as vibration absorbers [209], vibration damping devices [210–212], and so on, shown in Fig. 4.51.

For these MRE devices designed in the form of shear mode, the shear stiffness K which is an important parameter in structural vibration control and can be expressed as follows

$$K = K_0 + \Delta K = \frac{G_0 A}{h} + \frac{\Delta G A}{h} \quad (4.73)$$

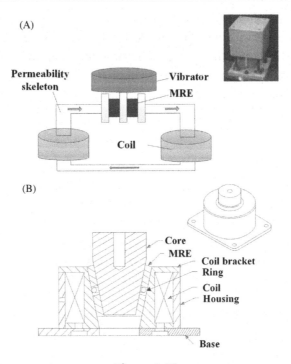

Figure 4.51
MRE devices. (A) MRE vibration absorber and (B) MRE vibration damping device.

where G_0 is the initial shear modulus of MRE and ΔG is the magnetic-induced shear modulus. In Eq. (4.73), ΔK is the increment of stiffness caused by the magnetic-induced shear modulus ΔG, which is an important index for MRE device design and affects the adjustability of MRE devices.

It can be seen from Fig. 4.51 that all kinds of MRE devices consist of two parts: MRE material and magnetic circuit structure.

1. The preparation of MRE: As a kind of MR materials, the magnetic-induced effect is an important index to measure the performance of MREs. So, the MRE materials used for designing devices should have a good magnetic-induced effect. As known, there are three main material components, including matrix, particles, and additives, are used to fabricate MREs. In order to improve the magnetic-induced effect of MRE, a lot of experiments have been carried out by using different materials, such as different matrix (silicone rubber, gelatin, resins, and so on), different magnetic particles (ferrite particles and carbonyl iron particles) and different additives [212–214]. On the other hand, as a kind of engineering materials, the MRE should have good mechanical properties to meet its small strain behavior.

2. Magnetic circuit structure: The magnetic circuit structure of MRE device would provide a uniform magnetic field for the MRE working area. The purpose of designing a magnetic structure is to construct a magnetic structure with low magnetic resistance, which can guide and focus the magnetic field on the MRE working area, and as far as possible to reduce magnetic energy losses in other areas.

4.7.3 Mathematical Models

So far, many researchers have theoretically and experimentally studied MRE devices under external magnetic field. In order to realize the calculation of structures incorporated with MRE devices, a mathematical model should be deduced to describe the performance of MREs under a magnetic field. Till now, some models were proposed in analyzing the field-dependent modulus of MREs based on simple chain mechanisms and particle dipole interactions, as mentioned in Section 4.7.1. These microphysical models [206–208] were proposed to calculate the magnetic-induced modulus of MRE. These models are deduced from the microaspect, and can be used to calculate the magnetic-induced shear modulus. However, the final purpose of proposing these models is to explain the macroscopic mechanical behavior of MRE and can be used to simulate force−displacement of MRE devices. Thus, the macromodel of MRE device should be proposed.

From these experimental results, we can see that the MRE material exhibits a feature that its modulus and damping capability are both magnetic field dependent, and the behavior of MRE without an external magnetic field processes the viscoelastic property. Generally speaking, the mechanical behaviors of MRE can be described as magnetic-induced viscoelastic property. Based on the viscoelastic theory, some parametric models were proposed to describe the magnetic-induced viscoelastic properties of MRE. A constitutive model was derived based on the viscoelasticity theory, in which four variables are adopted in the constitutive function to capture the viscoelastic characteristics of the matrix and the magnetic-induced effect [215]. Another four-parameter linear viscoelastic model was proposed, which could predict the dynamic mechanical property of MRE performance under various working conditions [216]. In addition, a fractional derivative magnetic viscoelastic parameter model was proposed to describe the magnetic viscoelastic mechanical behavior of MR elastomers under external magnetic field [217]. These macroparameter models are shown in Fig. 4.52.

It can be seen from Fig. 4.52 that several simple mechanical components, such as the spring element and so on, are combined together in the form of series or parallel. The physical interpretations of these mechanical elements in these models separately are the shear modulus of the matrix (spring element), the damping coefficient of MRE (Newton pot element), and the magnetic-induced shear modulus of MRE (nonlinear spring element). Then, the constitutive equations of these models were built, and take the fractional

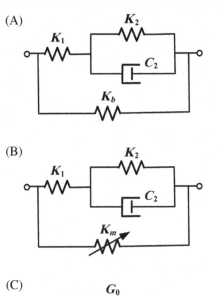

Figure 4.52
Macro parameter model for MRE. (A) Four-parameter viscoelastic model for MRE, (B) nonlinear four-parameter model for MRE, and (C) fractional derivative magnetic viscoelastic parameter.

derivative magnetic viscoelastic parameter model for example. Supposing the input strain is ε, while the output stress is σ, the constitutive equation can be expressed as follows

$$\sigma(t) = G_0 \varepsilon(t) + G_0 \tau^\alpha D^\alpha[\varepsilon(t)] + G_m(H)\varepsilon(t) + \mu G_m(H)\tau\dot{\varepsilon}(t) \tag{4.74}$$

where α is the fractional order. In order to get the complex modulus G^*, the Fourier transformation is undertaken for Eq. (4.74), then

$$G^* = \frac{\sigma(\omega)}{\varepsilon(\omega)} = G_0 + G_0 \tau^\alpha (i\omega)^\alpha + G_m(H) + i\mu G_m(H)\tau\omega = G_1(\omega) + iG_2(\omega) \tag{4.75}$$

where G_1 and G_2 are real and imaginary parts of the complex modulus, respectively. The $i^\alpha = \cos(\alpha\pi/2) + i\sin(\alpha\pi/2)$ is undertaken for Eq. (4.75), then the storage modulus G_1 and loss modulus G_2 can be obtained

$$G_1 = G_m(H) + G_0(1 + \tau^\alpha \omega^\alpha \cos\frac{\alpha\pi}{2}) \tag{4.76}$$

$$G_2 = \mu G_m(H)\omega\tau + G_0\tau^\alpha\omega^\alpha \sin\frac{\alpha\pi}{2} \tag{4.77}$$

and the loss factor can be calculated

$$\eta = \frac{G_2}{G_1} = \frac{\mu G_m(H)\omega + G_0\tau\omega^\alpha \sin(\alpha\pi/2)}{G_m(H) + G_0(1 + \tau\omega^\alpha \cos(\alpha\pi/2))} \tag{4.78}$$

where ω is the excitation frequency.

The parameters in the equations can be identified by using the experimental data, then the equivalent stiffness used in the structural vibration control analysis can be calculated by Eq. (4.73).

4.7.4 Analysis and Design Methods

4.7.4.1 MRE vibration absorber

The dynamic vibration absorber is a tuned spring-mass system, which reduces or eliminates the vibration of a harmonically excited system. The working principle of the dynamic vibration absorber, as shown in Fig. 4.53A, is added to the mass spring system on the vibration objects. In order to let the amplitude of the primary system be zero, the design working frequency of the vibration absorber should be close to the excitation frequency. The design working frequency is

$$f_a = \frac{1}{2\pi}\sqrt{\frac{K_2}{m_2}} \to f_e \tag{4.79}$$

From Eq. (4.79), we can see that the working frequency of the ordinary vibration absorbers is one fold or narrow brand. The working frequency of MRE vibration absorber can be changed in real time because the stiffness is controllable under external magnetic field.

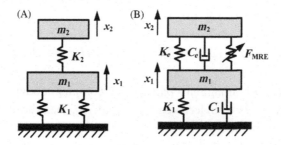

Figure 4.53
MRE device working principle. (A) Vibration absorber and (B) damping device.

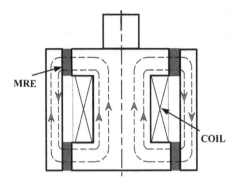

Figure 4.54
The magnetic circuit structure of MRE devices.

That is to say, the working frequency can be designed to trace the excitation frequency by changing the shear modulus of MRE. Therefore, combined with Eq. (4.73), the working frequencies can be designed by optimizing these parameters of A, h, and m.

4.7.4.2 MRE damping device

As a kind of semiactive vibration control device, the working principle of MRE is the same as that of MR damper, which can provide a "control force" F to the whole system. Fig. 4.53B is the working principle of the MRE vibration damping device, in which F can be expressed as follows

$$F_{\text{MRE}} = \Delta K \cdot s_T \quad (4.80)$$

where $\Delta K = dI^2 + eI + f$, d, e, f separately are the constants fitted by the experimental data, and s_T is the shear deformation.

Based on the design experience of MR damper, the elastic force F (control force), similar to the damping force of MR damper, is the most important index for designing MRE damping devices. From Eq. (4.73), we can see that $\Delta K = \Delta G A/h$ is the stiffness increment of MRE device produced by the magnetic-induced shear modulus.

From the microphysical model, we can see that the magnetic-induced shear modulus ΔG is mainly affected by the external magnetic field H when the MRE material is prepared. So, a reasonable magnetic circuit structure is needed to provide a uniform and enough magnetic fields for MRE working area, as seen in Fig. 4.54.

4.7.5 Tests and Engineering Applications

Since 1995 MREs were found, a series of the performance experiments of MREs were conducted and some of the MRE devices, such as, controllable stiffness of car bushing and adjustable vibration absorber were designed, which were used and applied for patents [218–220]. Shiga et al. [221] prepared two MRE samples using the silicon rubber as the matrix, then the mechanical properties of the MRE were studied with parallel plate shear model test. Jolly et al. [222] examined the mechanical response of elastomer composites subjected to magnetic fields, and these elastomer composites consist of carbonyl iron particles embedded within a molded elastomer matrix. The composite is subjected to a strong magnetic field during curing, which causes the iron particles to form columnar structures that are parallel to the applied field. Lokander and Stenberg [223] studied the ways to increase the MR effect, results showed that the absolute MR effect of isotropic MR rubber materials with large irregular iron particles was independent of the matrix material, and the relative MR effect could be increased by the addition of plasticizers. Of course, the MR effect can be increased by increasing the magnetic field, although the material saturation will emerge at large fields. Mysore [224] prepared two types of MRE with different concentrations, and circular and rectangular shapes having thicknesses from 6.35 mm to a maximum of 25.4 mm, and these samples were tested under quasi-static compression and quasi-static double lap shear. The results showed that the measured off-state shear modulus had large variations with increase in the thickness of the sample, and the measured shear modulus from the double lap shear test results, as well as the Young's modulus from the compression tests at zero-field, followed a logarithmic trend. An adaptive variable stiffness MRE absorber was developed by Dong et al. [225], and the working characteristics of this developed MRE absorber were also tested in a two-degree-of-freedom dynamic system. In addition, an active-damping-compensated MRE adaptive tuned vibration absorber was proposed by Gong et al. [226] and the dynamic properties and vibration attenuation performances were also experimentally investigated. Li et al. [210] designed a continuously variable stiffness MRE isolator used in vehicle seat suspension and the vibration control effect of the MRE isolator was evaluated. An isolation bearing with MRE used for structural vibration control in civil engineering was designed by Tu et al. [227].

Recently, for the deeper research on magnetic-characteristics and a wider range of applications of MRE, experimental researches on mechanical performance of MREs, especially on shear performance, have been studied, the results of which provide theoretical basis for the MRE devices with the shear working mode. The shear performances of MRE, that is, the dynamic viscoelastic properties under different magnetic fields, displacement amplitudes, and frequencies, were tested by the equipment, similarly as shown in Fig. 4.55. For example, the MRE fabricated with bromobutyl rubber and natural rubber was designed by Xu et al. [228] and the performances of the kind of MRE devices was tested, in which the test sample was manufactured in a style of a sandwich, as shown in Fig. 4.56.

These MRE samples were tested with respect to harmonic loadings in accordance with various displacement amplitudes (1, 2, and 4 mm) and frequencies (1, 2, 5, and 10 Hz). Originally, the experimental force displacement traces were recorded and only one cycle was taken for the calculations to ensure that the values represented only the steady state,

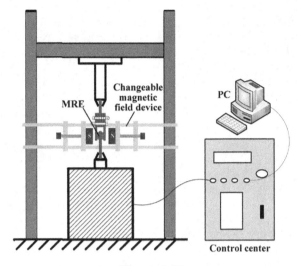

Figure 4.55
The MRE performance test system.

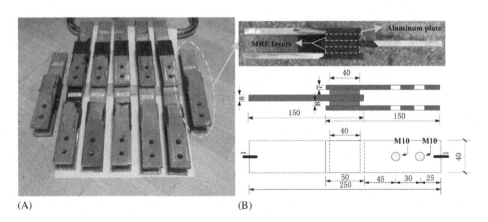

Figure 4.56
MRE test samples. (A) MRE sample photo and (B) structure diagram of the MRE samples.

and the test process is shown in Fig. 4.57. Figs. 4.58–4.60 show the force–displacement hysteresis curves of the two MREs under different currents, displacement amplitudes, and excitation frequencies.

Fig. 4.58 shows the force–displacement relationships of the two kinds of MRE samples (take NR70 and BIIR70 for example) at the constant displacement amplitude of 2 mm, frequency of 10 Hz but at various magnetic fields from 0 to 300 mT. From this figure, we can see that the hysteresis loops of the two kinds of MRE can shape ellipses, and the slope

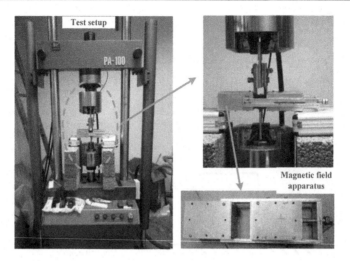

Figure 4.57
Experimental setup.

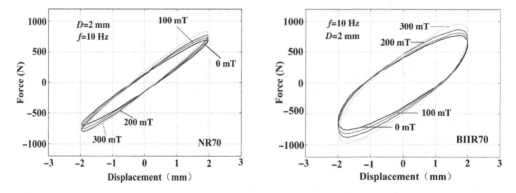

Figure 4.58
Force−displacement curves of MRE for different magnetic fields.

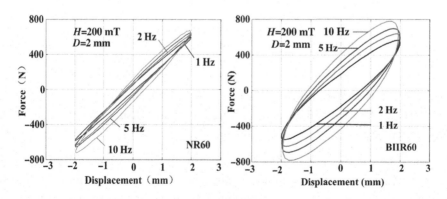

Figure 4.59
Force−displacement curves of MRE devices for different frequencies.

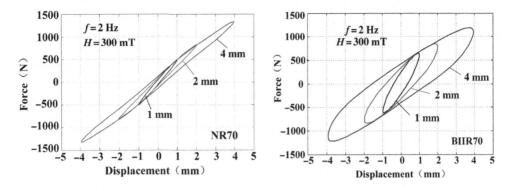

Figure 4.60
Force−displacement curves of MRE devices for various displacement amplitudes.

of the major axis of the elliptical loops varies with the magnetic field. The experimental force F increases with the increment of magnetic field strength values while the displacement amplitude is constant. This means that because of the changes in the material, a greater force is required to maintain the same level of displacement, that is to say, the stiffness of MREs varies with the magnetic fields. The areas of the elliptical loops increase slightly with the increment of the magnetic fields, which demonstrates that the damping capacity of the MRE samples is a function of the applied magnetic fields.

Fig. 4.59 shows the force−displacement curves of the MRE samples, NR60 and BIIR60, at the constant displacement amplitude of 2 mm, magnetic fields of 200 mT but at various frequencies from 1 to 10 Hz. It can be seen from these figures that the changes in slope of the main axis of hysteresis loops are not significant, whereas the loops area increases markedly, which demonstrates that the damping capacity of the MRE specimens is greatly influenced by the excitation frequencies.

Fig. 4.60 shows the force−displacement curves of NR70 and BIIR70 at the constant frequency of 2 Hz, the magnetic field of 200 mT but at various displacement amplitudes. It can be seen from Fig. 4.60 that the slopes of the main axis of these elliptical loops obviously change with the increment of the displacement amplitudes, which means that the stiffness of MRE specimens decreases when the displacement amplitudes increase from 1 to 4 mm.

Due to the controllable performance of MRE devices, they have been used in the automobile suspension system to reduce the deformation of the suspension system and increase the comfortableness [219]. However, the MRE devices used in the practical civil engineering projects have not been reported as far as the author concerned, and literatures are only limited in theoretical and experimental analysis. Along with the deepening of researches, the performance of MRE will be improved and may be used in the practical engineering project in the future.

CHAPTER 5

Design and Parameters Optimization on Intelligent Control Devices

Different kinds of intelligent control devices have been introduced in detail in the previous chapters, it can be seen that these devices can effectively dissipate the vibration energy. However, the parameters of these devices will definitely dominate the dissipation properties. Unreasonable parameters may even degrade the performance of these devices and cause economic waste, thus parameters optimization of these devices should be conducted. In this chapter, magnetorheological (MR) damper, MR elastomer, and the active mass damper (AMD) control system are taken as examples to introduce the design and parameters optimization process.

5.1 Design and Parameters Optimization on MR Damper

As discussed in Section 4.2, the MR damper can be mainly classified into four categories according to the stress pattern of MR fluids. Actually, the commonly used MR dampers in civil engineering structures are in flow mode (valve mode) or mix mode (combination of valve mode and direct shear mode). In this section, a multistage shear-valve mode (mix mode) MR damper is taken as an example to illustrate the design and parameters optimization process on MR damper.

5.1.1 Design on MR Damper

The designing of MR damper is a complex work, the process mainly includes materials selection, geometry design, and magnetic circuit design, and they will be introduced in detail in the following.

5.1.1.1 Materials selection

MR fluid directly dominates the property of MR dampers, and thus the fine rheological effect, antisettlement characteristics, and high magnetic saturation intensity are expected for MR fluid. In recent 20 years, Xu's research group [229] developed many MR fluids, which have fine antisettlement characteristics and low apparent viscosity in low frequency vibration environment. Properties of a representative MR fluid are listed in Table 5.1.

Table 5.1: Performance parameters of MR fluid

Type	MRFXZD08-01
Density (g/ml)	3.09
Solid content (%)	81.24
Flash point	>250°C
Apparent viscosity (30°C,500/s)	240 mPa·s
Operating temperature	−40−180°C
Coefficient of thermal expansion	$3.4 \times 10^{-4}(1/°C)$
Specific heat	1.36×10^5 J/(kg°C)
Magnetic saturation yield strength (1.0 T)	35 kPa

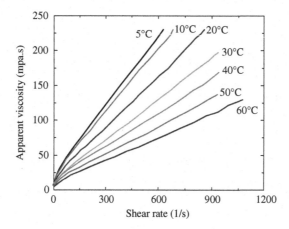

Figure 5.1
Apparent viscosity of MR fluid under different temperature.

Fig. 5.1 shows the apparent viscosity curves under different temperatures, and Fig. 5.2 shows the yield shear stress with respect to magnetic induction intensity.

Magnetic core in the magnetic circuit is used to increase magnetic induction, change the magnetic flux intensity, and reduce magnetic flux leakage. The materials usually used for magnetic core are electrical pure iron, iron−nickel alloy, iron−aluminum alloy, soft magnetic ferrite, and so on. Compared with the traditional dampers, the material of MR dampers should meet the requirement of mechanical properties as well as good permeability performance. The material for magnetic core should consider the following items.

High magnetic permeability, it will produce high magnetic induction intensity when only small current is inputted.

The area of magnetic hysteretic loop should be small, the coercive force and the magnetic hysteresis loss should be small.

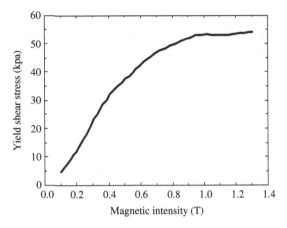

Figure 5.2
$\tau_y - B$ curve of MR fluid.

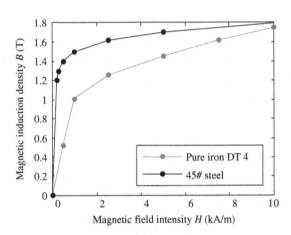

Figure 5.3
B−H curves of magnetic circuit material.

Good demagnetization ability, the magnetic field should quickly reduce to zero when the current is withdrawn.

The MR damper used in civil engineering structures consists of cylinder, piston, and the damping channel. The saturation induction density of the cylinder and piston should be higher than the magnetic field intensity of the MR fluid achieving magnetic saturation yield strength, so that the MR fluid can be fully utilized. What is more, the two parts are not only the parts of magnetic circuit, but also the main force delivery members of MR dampers. Therefore DT4 electrical pure iron and No. 45 steel, which have high magnetic permeability and high saturation induction density, are adopted for manufacturing the piston and the cylinder, respectively, whose $B-H$ curves can be seen in Fig. 5.3. It can be seen that the saturation induction densities of the DT4 electrical pure iron and No. 45 steel are higher than the magnetic field intensity when MR fluid achieves magnetic saturation yield strength, this shows that the selected materials are meeting the requirements of the goals.

The resistance and reactance of the coil directly dominate the quantity of current, and the reactance of the coil is also related to the reluctance of the magnetic circuit. In terms of excitation coil, the following items should be considered.

The coil should avoid being too hot or even destroyed under the maximum current.

Determine the turns of the coil according to the nominal diameter and section diameter of the coil.

The coil should not have leakage electricity caused by the motion of the piston or abrasion of the coil.

5.1.1.2 Design principle

When designing a MR damper, the controllable force and the dynamic range are usually two of the most important parameters in evaluating the overall performance of MR dampers [230,231]. The total damping force F produced by the shear-valve mode MR damper contains a controllable damping force F_τ, a plastic viscous force F_η, and a friction force F_f [232], as shown in Fig. 5.4.

$$F = F_\tau + F_\eta + F_f \tag{5.1}$$

In which, according to the parallel-plate Bingham model [233]

$$F_\tau = \left(2.07 + \frac{12Q\eta}{12Q\eta + 0.4wh^2\tau_y}\right)\frac{\tau_y L A_p}{h}\mathrm{sgn}(v_0). \tag{5.2}$$

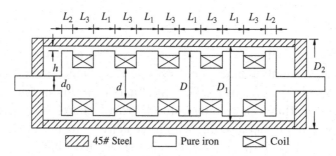

Figure 5.4
Schematic of the multistage shear-valve mode MR damper.

$$F_\eta = \left(1 + \frac{whv_0}{2Q}\right) \frac{12\eta Q L_t A_p}{wh^3}. \tag{5.3}$$

where $L = L_1 + L_2$ is the effective axial pole length. $L_t = L_1 + L_2 + L_3$ is the total axial pole length. w is the mean circumference of the damper's annular flow path $w = \pi(h/2 + D + h/2)$, and D is the diameter of the piston head. h is the height of the annular flow path between piston and cylinder. η and τ_y are the apparent viscosity without magnetic field and the yield shear strength of the MR fluid. A_p is the effective cross-sectional area of the piston head $A_p = \pi(D^2 - d_0^2)/4$, and d_0 is the diameter of the piston guide rod. Q is the volumetric flow rate $Q = v_0 \cdot A_p$. v_0 is the relative velocity between piston and cylinder, and sign function is used to consider the reciprocating of the piston.

According to Eq. (5.1), the total damping force can be decomposed into a controllable force F_τ and an uncontrollable force $F_{uc} = (F_{uc} = F_\eta + F_f)$. The uncontrollable force includes a plastic viscous force F_η and a mechanical friction force F_f. The dynamic range K is defined as the ratio between the controllable damping force F_τ and the uncontrollable force F_{uc}. If assume that $F_f = 0$, then

$$K = \frac{F_\tau}{F_{uc}} = \frac{F_\tau}{F_\eta + F_f} = \frac{F_\tau}{F_\eta}. \tag{5.4}$$

For a given shock absorption application engineering, the dynamic range K and the controllable damping force F_τ of MR dampers should be as large as possible, so that the designed MR damper can provide the engineering structure with the necessary damping force.

5.1.1.3 Geometry design

The task of geometry design is to choose an appropriate gap h, an effective active pole length L, and piston diameter d_0 to satisfy the design requirements of adjusting coefficient K and controllable force F_τ. These three parameters directly dominate the damping force provided by MR damper. The key steps are as follows:

- The objectives, namely the total damping force F and the adjusting coefficient K, should be determined according to the actual demand.
- Predetermine the following parameters: damping gap h, effective active pole length L, and piston diameter d_0. Computing the damping force and adjusting coefficient according to the above equations; if the total damping force F and the adjusting coefficient K do not meet the requirement, then modifying the parameters until the results meet the requirement.
- Check the strength of the element of the MR damper.

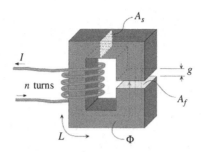

Figure 5.5
Simplified magnetic circuit of MR damper.

5.1.1.4 Magnetic circuit design

Magnetic circuit design is another important step to design a MR damper, magnetic circuit ohm's law is the basic formula to compute the magnetic induction intensity, and it is the basic principle for designing electromagnetic devices.

Usually, the magnetic circuit of the MR damper can be simplified as shown in Fig. 5.5. g is the air gap length. According to the magnetic circuit ohm's law, the following equation can be obtained.

$$NI = \Phi(R_g + R_f) = \Phi\left(\frac{l_1}{\mu_g A_g} + \frac{g}{\mu_f A_f}\right) \tag{5.5}$$

where R_g and R_f are the magnetic reluctance of the magnetic core and air gap. The magnetic motive force is mainly used to overcome the magnetic reluctance of the air gap due to that the air gap has a large magnetic reluctance and small magnetic permeability, while the magnetic coil has large magnetic permeability and small magnetic reluctance. Thus the length of the air gap should be small as far as possible.

The objective of magnetic circuit design is to determine the number of the coils, so that the magnetic induction density in the annular flow path generated by magnetic circuit is larger than the magnetic saturation yield strength of MR fluid.

5.1.2 Parameters Optimization on MR Damper

The parameters optimization on MR damper includes two parts. One is the geometric optimization, and the other is the magnetic circuit optimization. Based on the discussion in Section 5.1.1, the optimization process will be introduced as follows.

5.1.2.1 Geometric optimization

The geometry of the MR damper depicted in Fig. 5.4 is designed considering engineering and test requirements, the maximum stroke and the required damping force of the MR damper are predetermined as 50 mm and 200 KN. The objective in designing the damper is to obtain a larger dynamic range K with the damping force of 200 KN at the velocity of 100 mm/s. Therefore the geometry design process becomes the optimization problem of a constrained nonlinear multivariable function. The objective function, constraint condition, and design variables are as follows:

Object function: A larger dynamic range $K > 15$.

Constraint condition: Total damping force $F = 200$ KN at the velocity of 100 mm/s.

Design variables: Diameter of the piston head D, diameter of the piston rod d_0, effective axial pole length L, total axial pole length L_t, and gap size h.

According to the objective function and constraint condition, the problem has been solved by properly adjusting the design variables using the function of *fmincon* in MATLAB. In this investigation, the designed variables are determined as $D = 156$ mm, $d_0 = 80$ mm, $L = 4 \times L_1 + 2 \times L_2 = 4 \times 50 + 2 \times 25 = 250$ mm, $L_t = 500$ mm, and $h = 2$ mm. In addition, main geometric parameters of the MR damper are shown in Table 5.2. Substituting the parameters into Eqs. (5.1)–(5.4), then

$$
\begin{aligned}
F_\tau &= \left(2.07 + \frac{12Q\eta}{12Q\eta + 0.4wh^2\tau_y}\right) \frac{\tau_y L A_p}{h} \\
&= \left(2.07 + \frac{12 \times 0.00141 \times 0.24}{12 \times 0.00141 \times 0.24 + 0.4 \times 0.49 \times 0.002^2 \times 35}\right) \times \frac{35 \times 0.25 \times 0.0141}{0.002} \\
&= 188.97 \text{ KN}
\end{aligned}
$$
(5.6)

$$
\begin{aligned}
F_\eta &= \left(1 + \frac{whv_0}{2Q}\right) \frac{12\eta Q L_t A_p}{wh^3} \\
&= \left(1 + \frac{0.49 \times 0.002 \times 0.1}{2 \times 0.00141}\right) \times \frac{12 \times 0.24 \times 0.00141 \times 0.5 \times 0.0141}{0.49 \times 0.002^3} \\
&= 7.56 \text{ KN}
\end{aligned}
$$
(5.7)

Table 5.2: Main geometric parameters of MR damper

Stroke (mm)	± 50
External diameter of cylinder (mm)	194
Internal diameter of cylinder (mm)	160
Diameter of the piston rod (mm)	80
Effective length of the piston (mm)	4 × 50 + 2 × 25 = 250
Damping gap (mm)	2
Trench diameter of coil (mm)	110

$$F = F_\tau + F_\eta = 188.97 + 7.56 = 196.53 \text{ KN} \tag{5.8}$$

$$K = \frac{F_\tau}{F_\eta} = \frac{188.97}{7.56} = 25 \tag{5.9}$$

As a result, the dynamic range is 25, the controllable damping force F_τ is 189 KN and the maximum damping force is 197 KN, these parameters can satisfy the design requirements completely.

5.1.2.2 Magnetic circuit optimization

Magnetic circuit directly determines the size of the magnetic induction density in the annular flow path. An optimal design of the magnetic circuit is required to maximize magnetic induction density in the annular flow path and minimize the energy loss in steel flux conduct and regions of nonworking areas. Here only one magnetic circuit will be designed because each stage of the magnetic circuit structure is the same as shown in Fig. 5.6.

According to Ohm's law [234], the following equation can be obtained

$$NI = \Phi \cdot (R_h + R_f) \tag{5.10}$$

where Φ is the magnetic flux, R_f and R_h are the total iron core reluctance and annular flow path reluctance, respectively. I is the current, N is the number of the coils, $N = m \times n$, m is the layer of the coil, n and d_w are the number of the coils at each layer and the diameter of the wire, respectively, as shown in Fig. 5.6.

Because magnetic circuit consists mainly of magnetic core, yoke, out cylinder, and annular flow path, the total iron core magnetic reluctance R_f is considered to be composed of magnetic core reluctance R_1, yoke reluctance R_2, and cylinder reluctance R_3. The relative permeability of No. 45 steel (μ_{steel}), MR fluid (μ_{MRF}), electrical pure iron DT4 (μ_{iron}), and

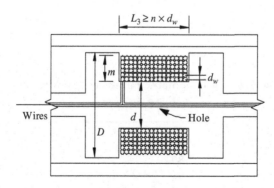

Figure 5.6
Magnetic circuit assembly.

copper wire (μ_{coil}) are 1000, 4, 1600, and 1, respectively. Then, considering the geometric parameters that have been determined in the geometry design, the magnetic reluctance R_1, R_2, R_3, and R_h can be calculated according to Ohm's law and the pole surface of concentric cylinders gap permeance formula.

$$R_1 = \frac{4(L_1 + L_2)}{\pi \mu_{steel} d^2} = \frac{4 \times (0.05 + 0.025)}{\pi \times 1600 \times 4\pi \times 10^{-7} \times 0.11^2} = 3.93 \times 10^3 \quad (5.11)$$

$$R_2 = \frac{\ln(D/d)}{2\pi \mu_{iron} L_2} = \frac{\ln(0.156/0.11)}{2\pi \times 1600 \times 4\pi \times 10^{-7} \times 0.025} = 1.11 \times 10^3 \quad (5.12)$$

$$R_3 = \frac{4(L_1 + L_2)}{\pi \mu_{steel}(D_2^2 - D_1^2)} = \frac{4 \times (0.05 + 0.025)}{\pi \times 1000 \times 4\pi \times 10^{-7} \times (0.194^2 - 0.16^2)} = 6.31 \times 10^3 \quad (5.13)$$

$$R_h = \frac{\ln(D_1/D)}{2\pi \mu_{MRF} L_2} = \frac{\ln(0.16/0.156)}{2\pi \times 4 \times 4\pi \times 10^{-7} \times 0.025} = 3.21 \times 10^4 \quad (5.14)$$

Then, the total magnetic reluctance is calculated by the following equations

$$R_m = R_1 + 2R_2 + R_3 + 2R_h = 7.666 \times 10^4 \quad (5.15)$$

According to magnetic saturation yield strength of MR fluid, take 1.0 T as the design value of the magnetic induction density in the annular flow path, then the magnetic flux in the magnetic circuit can be calculated as follows:

$$\Phi = B \times S = 1.0 \times \pi \times \left(\frac{0.160 + 0.156}{2}\right) \times 0.025 = 0.0124 \quad (5.16)$$

From Eq. (5.10), the current can be estimated as follows:

$$I = \frac{\Phi \cdot (R_h + R_f)}{N} = \frac{\Phi \cdot R_m}{N} \quad (5.17)$$

Then, the optimization problem with the magnetic circuit design is converted into searching for the design variables of a constrained nonlinear multivariable function. The objective function, constraint condition, and design variables are as follows:

Object function: $I \leq 3$ A (maximum current of the direct current power supply).

Constraint condition: $NI \geq \Phi \cdot R_m = 0.0124 \times 7.666 \times 10^4 = 950$ AT, $L_3 \geq n \times d_w$, $\frac{D-d}{2} \geq m \times d_w$.

Design variables: m is the layer of the coil, n is the number of the coil at every layer, and d_w is the diameter of the coil.

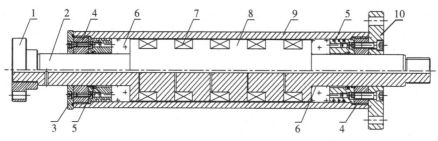

1. Upper connection flange
2. Guide rod of the piston
3. Press plate of the cylinder
4. Guides leeve
5. Seal installation
6. MR fluid
7. Coils
8. Piston
9. Cylinder
10. Bottom connection flange

Figure 5.7
Assembly diagram of the MR damper in this study.

The problem can be solved by properly adjusting the design variables according to the objective function and the constraint conditions by using the function of *fmincon* in MATLAB. In this paper, the designed variables are $m = 20$, $n = 42$, $d_w = 1.0$ mm, and the corresponding variables are $N = m \times n = 840$, $I = 1.13$ A, $L_3 = 50 \geq 42 \times 1 = 42$, $\frac{D-d}{2} = \frac{156-110}{2} = 23 \geq m \times d_w = 20 \times 1 = 20$.

Based on the above design and optimization process, the MR damper is manufactured and assembled according to the results of geometry design and magnetic circuit design, as shown in Fig. 5.7.

5.2 Design and Parameters Optimization of MRE Device

In this section, the design and parameters optimization process of magnetorheological elastomer (MRE) devices will be discussed, the MRE device mentioned in Section 4.7.5 [210] will be taken as an example for illustration. The working performance of the MRE device depends on the value of the control force, which can be calculated by Eq. (4.80) in Section 4.7. In addition, it can be seen from Eqs. (4.70) to (4.73) that the magnetic-induced modulus is the main factor which influences the control force, while the magnetic circuit directly dominates the magnetic field and then affects the magnitude of the magnetic-induced modulus. Thus it is essential to optimize the magnetic circuit to obtain a larger magnetic-induced modulus in MRE device design.

5.2.1 Parameters Optimization for Magnetic Circuit

For the MRE device introduced in Section 4.7.5 [210], the geometry configuration parameters of the magnetic circuit structure of the designed MRE device are shown in

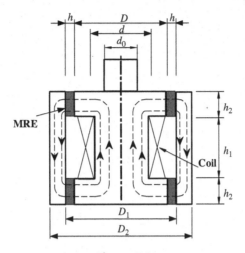

Figure 5.8
The parameters of the magnetic circuit structure of the MRE device.

Fig. 5.8. The aim of parameter design is to design a magnetic circuit with low magnetic resistance, which can make the magnetic flux focus on the MRE working zone and avoid leakage flux occurring in other zones. The magnetic circuit consists of magnetic core (piston) and magnetic yoke (sleeve), so the magnetic resistance of each part should be calculated, which relates to the size of the member.

Based on the Ohm's law, the relationship between magnetic flux Φ and magneto-motive F can be expressed as

$$\Phi = \frac{F}{R_m} \tag{5.18}$$

$$F = NI \tag{5.19}$$

$$\Phi = \int_S B dS \tag{5.20}$$

$$R_m = \frac{l}{\mu_0 \mu S} \tag{5.21}$$

where R_m is magnetic resistance, N is number of coil cycles, I is field current, B is magnetic flux density, S is the area of the magnetic circuit, l is the length of the magnetic circuit, and μ is relative permeability. But for the specific calculation, the magnetic circuit can be simplified and shown in Fig. 5.9. Then, combined with the concentric cylindrical surface gap permanence formulas, the magnetic resistance of magnetic core, magnetic yoke, sleeve, and MRE can be calculated.

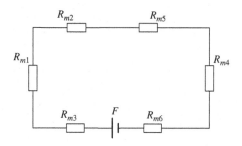

Figure 5.9
The simplified magnetic circuit of the MRE device.

The magnetic resistance R_{m1}, R_{m2}, and R_{m3} of magnetic core can be expressed as

$$R_{m1} = \frac{l_1}{\mu_0 \mu_1 S_1} = \frac{4(h_1 + h_2)}{\mu_0 \mu_1 \pi d^2} \tag{5.22}$$

$$R_{m2} = R_{m3} = \int_d^D \frac{1}{2\pi \mu_0 \mu_2 l_2 h_2} dl = \frac{\ln(D/d)}{2\pi \mu_0 \mu_2 h_2} \tag{5.23}$$

The magnetic resistance R_{m4} of sleeve can be expressed as

$$R_{m4} = \frac{l_4}{\mu_0 \mu_4 S_4} = \frac{4(h_1 + h_2)}{\mu_0 \mu_4 \pi (D_2^2 - D_1^2)} \tag{5.24}$$

The magnetic resistance R_{m5} and R_{m6} can be expressed as

$$R_{m5} = R_{m6} = \frac{l_5}{\mu_0 \mu_5 S_5} = \frac{\ln(D_1/D)}{2\mu_0 \mu_5 \pi h_2} \tag{5.25}$$

The magnetic resistance R_m of the whole magnetic circuit can be expressed as

$$R_m = R_{m1} + 2R_{m2} + R_{m4} + 2R_{m5} \tag{5.26}$$

Here, the relative permeability of these magnetic materials of the MRE device can be got from the manufacturer, $\mu_{steel} = 1000$ (the No. 45 steel), $\mu_{iron} = 1600$ (DT$_4$ electric pure iron), $\mu_{coil} = 1$ (copper interconnects), and the permeability of vacuum is $\mu_0 = 4\pi \times 10^{-7}$ Wb/A·m. However, the relative permeability of MRE cannot be given directly, so the important work is to analyze the permeability of MRE.

It can be seen in the references [60–62] that MRE is fabricated with the mixture of carbonyl iron powder particles and rubber-like elastomers, and theoretical analysis method is difficult to compute the relative permeability due to the nonlinear characteristics.

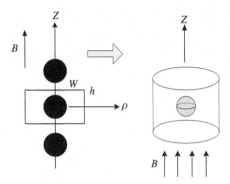

Figure 5.10
The schematic of element body.

Thus the electromagnetic finite element analysis method is used to simulate the magnetic field distribution under uniform magnetic field and calculate the relative permeability of MRE. One particle with the surrounded elastomers is detached as an element body which is shown in Fig. 5.10. The finite element model of the element body is built with the software ANSOFT Maxwell and the half section of the model along the cylinder axis is shown in Fig. 5.11.

Under the uniform magnetic field, the magnetic induction intensity distribution and magnetic force line distribution are calculated and shown in Fig. 5.12.

The relative permeability usually can be calculated by Eq. (5.27), where B_0 and H_0 separately are the average magnetic field intensity and the average magnetic induction intensity, respectively, which can be calculated by Eqs. (5.28) and (5.29). Therefore the calculated relative permeability of MRE can be obtained when the particles reach saturated magnetic intensity, which is about $u_r = 2.34$.

$$\mu_r = \frac{B_0}{\mu_0 H_0} \tag{5.27}$$

$$H_0 = \sum_{i=1}^{n} H_{z,i} V_i / \sum_{i=1}^{n} V_i \tag{5.28}$$

$$B_0 = \sum_{i=1}^{n} B_{z,i} V_i / \sum_{i=1}^{n} V_i \tag{5.29}$$

According to the magnetic flux conservation law, the magnetic circuit magnetic flux of all parts should be equal, i.e., $\Phi_1 = \Phi_2 = \Phi_3 = \Phi_4 = \Phi_5 = \Phi_6$. The magnetic induction intensity can be calculated by $\Phi = B \cdot S$. So, according to the relationship between the size of each

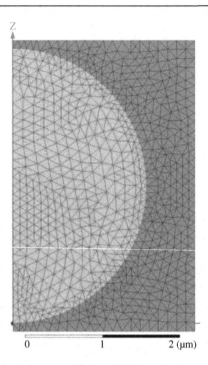

Figure 5.11
The finite element model.

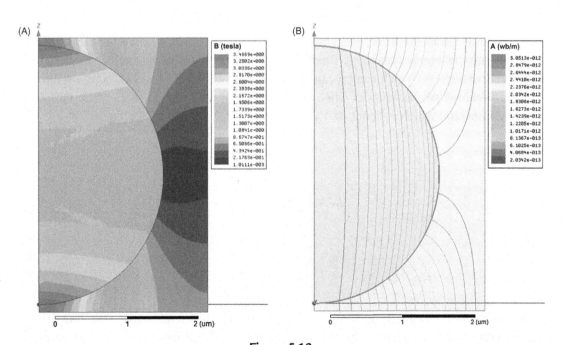

Figure 5.12
The calculation results. (A) The magnetic induction intensity distribution and (B) the magnetic field line distribution.

part of the magnetic circuit and magnetic resistance, in order to get better utilization of the magnetic circuit by optimizing the parameters, each part of the magnetic circuit should reach magnetic saturation one after another, and the difference of the magnetic induction intensity at magnetic saturation between each part is small. Based on the above principle, the size of each part of the magnetic circuit can be optimized, and the optimization result is as follows: $d = 38$ mm, $D = 58$ mm, $D_1 = 70$ mm, $D_2 = 82$ mm, $h = 6$ mm, $h_1 = 100$ mm, $h_2 = 10$ mm, and $D/d \approx 1.53$, $h_1/d \approx 2.11$, $(D_2 - D_1)/2(h_1 + h_2) \approx 0.055$. Substituting the above parameters into Eqs. (5.18)–(5.26), the magnetic induction intensity of each part of the magnetic circuit can be calculated and shown in Table 5.3.

It can be seen from Table 5.3 that when the MRE working zone reaches magnetic saturation, the other parts of the magnetic circuit basically reach magnetic saturation, which demonstrates that the design of the whole magnetic circuit is reasonable.

5.2.2 Magnetic Circuit FEM Simulation

It is very difficult to get the accurate analytical solution for complex magnetic circuit structures, while some numerical methods, such as Finite Element Method (FEM), are commonly used to solve this problem. Therefore the software ANSOFT Maxwell is chosen to analyze the magnetic field distribution of the MRE device.

According to Section 5.2.2, the finite element model of the MRE device is built and shown in Fig. 5.13, and the calculating result is shown in Fig. 5.14. It can be seen that the magnetic induction of the iron core is about 1.25–1.56 T, the sleeve is about 1.04–1.38 T and the MRE is about 0.53–0.73 T, which is lower than the calculation results. The reason is that the $B-H$ curve of materials is considered to be nonlinear in finite analysis but linear in the theoretical calculation. So the parameters should be adjusted to be in coincidence with the actual materials. Finally, the sizes of the iron core and sleeve are changed as: $d = 48$ mm, $D = 68$ mm, $D_1 = 80$ mm, and $D_2 = 96$ mm. The FEM model is rebuilt according to the changed parameters, and the analysis results are shown in Fig. 5.15.

It can be seen from Fig. 5.15 that the magnetic induction of iron core is about 1.32–1.65 T, the sleeve is about 1.1–1.32 T and the MRE is about 0.77–0.99 T, which are close to the

Table 5.3: The magnetic induction intensity of each part of the magnetic circuit

	R_{m1}	R_{m2}	R_{m3}	R_{m4}	R_{m5}	R_{m6}
Working zone (T)	1.5	1.5	1.5	1.2	0.85	0.85
Magnetic resistance (10^3/H)	48.2	3.35	3.35	61.1	1018	1018
Magnetic circuit area (m^2/10^3)	1.1	1.5	1.5	1.4	2.0	2.0
Magnetic induction intensity (T)	1.507	1.133	1.133	1.193	0.850	0.850
Magnetic saturation order	1	4		3	2	

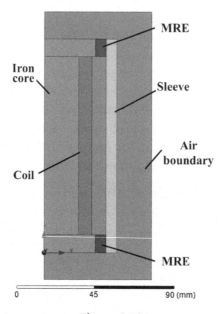

Figure 5.13
The finite analysis model.

calculation results. The detailed results of the two methods are shown in Table 5.4. It can be seen from Table 5.4 that the calculation results of the two methods match well, and there is a linear relationship between magnetic strength B and electric current I when the current is small (no more than 1.6 A). The above analysis results demonstrate that the magnetic circuit design of MRE device can be implemented with the simplified magnetic circuit method, which is based on the basic magnetic conductance theory. In other words, the simplified magnetic circuit method can be used for the preliminary design of MRE device. Then the FEM is used for further analysis of the magnetic circuit.

5.3 Design and Parameters Optimization on Active Control

Some active control systems, including active tendon system (ATS) and AMD, have been introduced in the Chapter 3, Active Intelligent Control, and the AMD control system is taken as an example to introduce the design and parameters optimization of active control systems in this section. The AMD control system consists of AMD system, controllers, and sensors, so the designed parameters involves three aspects: AMD system parameters, including the mass, damping, and stiffness of auxiliary mass; control parameters, including the maximum driving force and stroke of AMD; and placements of sensors and actuators. Here, some common design methods and corresponding parameters optimization of the AMD control system will be introduced in detail.

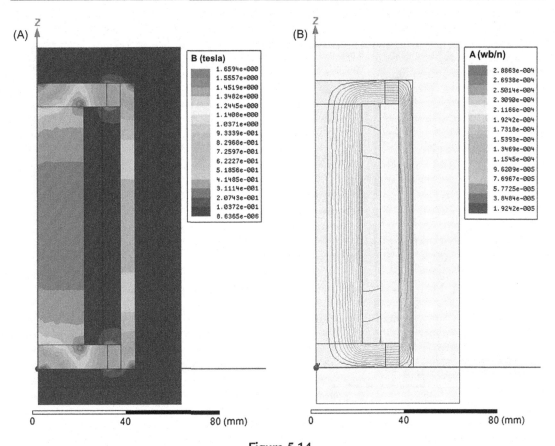

Figure 5.14
The calculation results (I = 2.5 A). (A) The distribution of magnetic induction and (B) the distribution of magnetic force line.

5.3.1 Design and Parameters Optimization Based on Feedback Gain

Sensors and actuators are typically placed on the position with the maximum structural vibration response, so the placement of sensors and actuators is not discussed. Nishimura et al. proposed a control strategy for AMD, the feedback gain and the AMD system parameters can be optimized simultaneously [235]. Neglecting the damping of the primary structure, Eqs. (3.3) and (3.4) can be rewritten as:

$$m\ddot{x}(t) + kx(t) = p(t) + c_a\dot{x}_a(t) + k_a x_a(t) - f_a(t) \tag{5.30}$$

$$m_a(\ddot{x}_a(t) + \ddot{x}(t)) + c_a\dot{x}_a(t) + k_a x_a(t) = f_a(t) \tag{5.31}$$

where $m_a\ddot{x}_a(t)$ is an active control force on the controlled structure, the control efficiency actually derived from the damping effect of tuned mass damper (TMD) on the primary

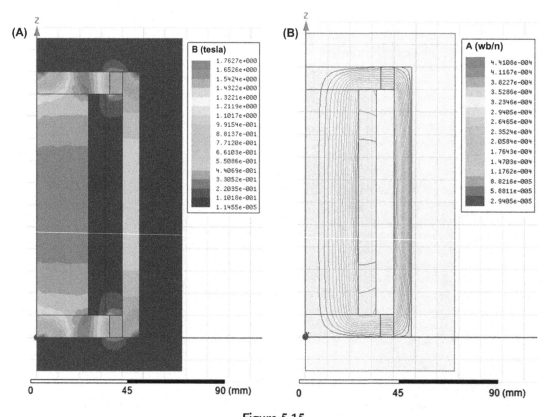

Figure 5.15
The calculation results (I = 2.5 A). (A) The distribution of magnetic induction and (B) the distribution of magnetic force line.

Table 5.4: The FEM and theoretical results of magnetic induction intensity at MRE working zone

Electric Current	0.2 A	0.4 A	0.6 A	0.8 A	1.0 A	1.2 A	1.4 A	1.6 A	1.8 A
Theoretical results (B/T)	0.069	0.139	0.208	0.278	0.347	0.416	0.486	0.555	0.625
FEM results (B/T)	0.071	0.135	0.207	0.281	0.352	0.420	0.495	0.572	0.652
Deviation (%)	2.80	−2.81	−0.58	1.28	1.39	0.81	1.82	2.90	4.20

structure and the output force of the actuator. The feedback gain of the proposed algorithm is linear to the response acceleration of the primal system, then the driving force can be expressed as

$$f_a(t) = -G\ddot{x}(t) = -mg\ddot{x}(t) \tag{5.32}$$

where G is the feedback gain and g is the normalized feedback gain. Introducing parameters $f = \frac{w}{w_0}$, $\beta = \frac{w_a}{w_0}$, $\mu = \frac{m_a}{m}$, and $\xi_a = \frac{c_a}{2m_a w_a}$, the frequency spectra response of the controlled structure is expressed as

$$|H(f)|^2 = \frac{(\beta^2 - f^2)^2 + (2\xi_a \beta f)^2}{[(1-g)f^4 - (1+\beta^2+\mu\beta^2)f^2 + \beta^2]^2 + [2\xi_a \beta f(1-f^2-\mu f^2)]^2} \quad (5.33)$$

The above response function of the controlled structure has two locked frequencies, where the corresponding responses are not influenced by the damping of the auxiliary mass. This phenomenon is similar to the case of TMD, so the optimization process of the TMD parameters can be conducted in the case of AMD. The final locked response value at the two locked points is given as

$$\alpha_{\max} = \sqrt{\frac{2 + \mu - g_{opt}}{\mu + g_{opt}}} \quad (5.34)$$

Hence the optimal feedback gain and the optimal frequency ratio are given as

$$g_{opt} = \frac{2 + \mu - \mu \alpha_{\max}^2}{\alpha_{\max}^2 + 1} \quad (5.35)$$

$$\beta_{opt} = \sqrt{\frac{\alpha_{\max}^2 - 1}{(\mu+1)(\alpha_{\max}^2+1)}} = \sqrt{\frac{1 - g_{opt}}{1 + \mu}} \quad (5.36)$$

The damping of auxiliary mass determines the peak of the response function, and the optimal damping should satisfy the condition that the peaks appear under the locked points. The approximated value of the optimal damping is given by the corresponding equation in TMD, and the validity of the equation is checked by various combinations of the parameters.

$$\xi_{opt} = \sqrt{\frac{3}{4(\alpha_{\max}^2+1)}} = \sqrt{\frac{3(\mu + g_{opt})}{8(1+\mu)}} \quad (5.37)$$

When the control effect and the value of the auxiliary mass are given, then the value of $\alpha_{\max}$, the optimal feedback gain, the optimal frequency ratio and the optimal damping can be calculated. It can be found that the feedback belongs to acceleration feedback. Chang and Yang adopted velocity feedback and complete feedback to design and optimize the feedback gain and AMD system parameters [236].

5.3.2 Design and Parameters Optimization Based on Minimum Energy Principle

As mentioned in Section 3.2, the active control force applied to the structure is not equivalent to the driving force derived from the actuator. Based on the difference, Liu et al.

proposed a design method of AMD control system, in which the active control force is determined firstly, and the next determination is about the driving force [237]. The equations of motion of the AMD control system can be expressed as

$$[M]\{\ddot{x}(t)\} + [C]\{\dot{x}(t)\} + [K]\{x(t)\} = \{P(t)\} + [B]\{f_d(t)\} \tag{5.38}$$

$$\{f_d(t)\} = [C_a]\{\dot{x}_a(t)\} + [K_a]\{x_a(t)\} - \{f_a(t)\} = -[M_a]\{\ddot{x}_{aa}(t)\} \tag{5.39}$$

where $\{f_d(t)\}$ is the active control force applied to the structure and $\{f_a(t)\}$ is the driving force derived from the actuator. In order to use active control algorithm to conduct the active control, Eq. (5.38) is rewritten as the state-space equation, and the performance index is defined as follows:

$$J = \int_0^\infty ((Z(t))^T[Q]\{Z(t)\} + \{f_d(t)\}^T \eta [R]\{f_d(t)\}) \, dt. \tag{5.40}$$

where η is the weight coefficient that determines the control effect of structural vibration control, and the control force is calculated according to linear quadratic regulator (LQR) algorithm. In structural vibration control, the active control force is always written as the combination of the gain matrix and the structural state, then the driving force can be rewritten as

$$\{f_a(t)\} = [C_a]\{\dot{x}_a(t)\} + [K_a]\{x_a(t)\} + [G]\{Z(t)\} \tag{5.41}$$

It can be seen from the above equation that the certain control effect must correspond to certain control force under a given active control algorithm. In other words, the provided control effect determines the control force, and the driving force is not determined due to the presence of AMD system parameters. The driving force can be calculated based on the relation between $\{f_d(t)\}$ and $\{f_a(t)\}$, but the selection of AMD system parameters is an important problem affecting the driving force. Obviously, the reasonable selection of parameters about AMD system can minimize the external input energy under a given vibration control effect, so the work done by the driving force is defined as the objective function

$$\begin{aligned} J &= \int_0^{t_f} (\{f_a(t)\}\{\dot{x}_a(t)\}) \, dt \\ &= \int_0^{t_f} (([C_a]\{\dot{x}_a(t)\} + [K_a]\{x_a(t)\} + [G]\{Z(t)\})\{\dot{x}_a(t)\}) \, dt \end{aligned} \tag{5.42}$$

$$x_a(t) = \int_0^t \int_0^t (-[M_a]^{-1}\{f_d(\tau)\} - [B]^T\{\ddot{x}(\tau)\}) \, d\tau \, d\tau \tag{5.43}$$

$$\dot{x}_a(t) = \int_0^t (-[M_a]^{-1}\{f_d(\tau)\} - [B]^T\{\ddot{x}(\tau)\}) \, d\tau \tag{5.44}$$

When the external excitation and the control effect are determined, the responses of the controlled structure and the active control force can be calculated. The basic parameters in objective function J are $[M_a]$, $[K_a]$, and $[C_a]$, which can be calculated by some numerical calculations. However, the selection of AMD system parameters is influenced by the external excitation, as a result the robustness is not good.

5.3.3 Design and Parameters Optimization Based on Fail-Safe Reliability

As described in Section 3.2, the design of control parameters of AMD control system is conducted after the optimization of the TMD parameters optimization, which can achieve a good control effect when the AMD fails to work or is not in a work state. For the parameters optimization of TMD, the mass of auxiliary mass can be determined according to the control effect and the practical engineering, the stiffness and damping of auxiliary mass can be obtained by the mature theoretical formula about optimal frequency ratio and damping ratio, such as the formula in Section 3.2 proposed by Tsai and Lin [73]. The driving force is calculated using active control algorithm, such as LQR algorithm described in Section 3.2.3.

It is worth noting that the weighting factors affect the control effect, the maximum driving force and the maximum stroke. The above indicators can be in a balanced state by optimizing the weighting factors. The detailed operation is to select reasonable performance index and define the range of parameters, such as stroke limits and maximum drive capability of AMD, then the numerical computation is adopted to calculate the reasonable weight factors.

$$[Q] = \alpha \begin{bmatrix} [K] & 0 \\ 0 & [M] \end{bmatrix} \cdot [R] = \beta [I] \tag{5.45}$$

where α and β are weighting factors. Although this design method has a high reliability of fail-safe, the control effect obtained is not optimal and economical, because the parameters are not based on global optimization.

CHAPTER 6
Design and Study on Intelligent Controller

In Chapter 2, Intelligent Control Strategies, different kinds of control strategies for intelligent vibration control are described. These control strategies are applied to the real structures by an intelligent controller in real time. In this chapter, the hardware design and experimental study for intelligent controller will be introduced. Taking the intelligent controller as an example, in 2004, Lord Corporation and Motorola Incorporation et al. [238] designed controllers of magnetorheological (MR) dampers in the automobile field. In 2007, Liu [239] presented a controllable current amplifier for MR damper. In 2012, Guo et al. [240] developed a single-chip microprocessor for MR damper coupling sensing and control.

The controller should include two parts: the sensing system and the control system. That is, the controller can collect status signals of the structures, and at the same time it can produce control signals of MR dampers according to status signals.

6.1 Design of Intelligent Controller

An intelligent controller for MR dampers coupling sensing and control functions is designed in accordance with the following principles: (1) the sensors collect the responses of the controlled structure and transmit them to the microcontroller in real time, which can be acceleration responses, displacement responses, velocity responses or stress responses, and so on; (2) the control current signals of MR dampers are calculated based on the collected responses of the controlled structure by microcontroller in which the control strategies are written in advance; (3) the control currents are applied to MR dampers through the pulse-width modulation (PWM) technology, and then MR dampers will produce different damping force according to different control currents to mitigate the structural vibration responses, as shown in Fig. 6.1. In the following section, the design method of the controller based on structural acceleration responses will be introduced in detail.

6.1.1 The Design of the Acceleration Responses Collection

In the process of the vibration mitigation control of structures with MR dampers, the structural acceleration response is usually considered as one of the most important parameters, which will clearly show the vibration condition of the structure, and it is also

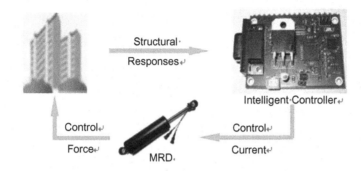

Figure 6.1
The MR intelligent controller.

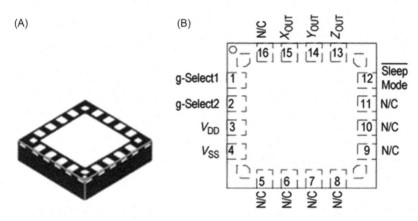

Figure 6.2
MMA7260QT. (A) Bottom view and (B) pin connections.

very useful to determine control currents of MR dampers, so the real-time acceleration responses of the controlled structure are adopted as the input of the control system. A three axes low-g micromachined accelerometer, MMA7260QT (as shown in Fig. 6.2), is used to collect the real-time acceleration responses of the structure.

The MMA7260QT [241] is a low-cost capacitive micromachined accelerometer. It features signal conditioning, a one-pole low-pass filter, temperature compensation, and g-Select, which allows for the selection among four sensitivities. Zero-g offset full scale span and filter cut-off are factory set and require no external devices. It includes a Sleep Mode that makes it ideal for handheld battery powered electronics. The device can measure both positive and negative accelerations, and transform acceleration of the voltage signal. With no input acceleration, the output is at mid-supply. For positive acceleration, the output will increase above $V_{DD}/2$ (V_{DD} is the supply voltage). For negative acceleration, the output

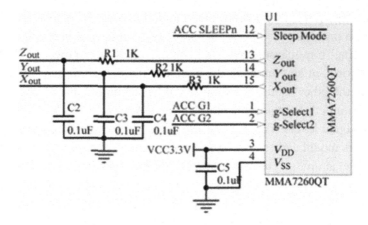

Figure 6.3
The connection diagram of MMA7260QT.

will decrease below $V_{DD}/2$. So the relationship between the acceleration input and the voltage output of the MMA7260QT is

$$V_{out} = \frac{V_{DD}}{8} A_{cc} + \frac{V_{DD}}{2} \tag{6.1}$$

where A_{cc} is the acceleration input and V_{out} is the voltage output. In this paper, according to the MMA7260QT's features, it is used to collect the three-way acceleration responses of the structure in real-time, and then the voltage output signal will be sent to the microcontroller. Fig. 6.3 shows the connection diagram of MMA7260QT.

6.1.2 The Design of the Microcontroller

The microcontroller chip is the core component of the MR damper intelligent controller. Its main function is to generate the control current signal of the MR damper based on the real-time vibration acceleration responses of the controlled system collected by the acceleration sensor. A eight-bit ATmega16 microcontroller is used to generate the control current signal to the MR dampers by using the PWM technology.

6.1.2.1 The PWM technology

In the process of controlling the MR dampers, the control currents of MR dampers will change from 0 to 3 A, but usually the circuit board only can withstand the maximum current of 1 A, that is, the controller cannot provide currents to MR dampers directly. In order to solve this problem, the PWM technology is used to control the power source and realize the control current supply of MR dampers.

The PWM technology [242] is a powerful technique for controlling analog circuits with digital outputs of a processor. The PWM technology is employed in a wide variety of applications, ranging from measurement and communications to power control and conversion. As known, at any moment, the output of the digital circuit only has two states, ON and OFF. On the other hand, an analog signal has a continuously varying value, with infinite resolution in both time and magnitude. The PWM technology is a way of digitally encoding analog signal levels. Through the use of high-resolution counters, the duty cycle of a square wave is modulated to encode a specific analog signal level. The PWM signal is still digital because, at any given instant of time, the full Direct Current (DC) supply is either fully on or fully off. The voltage or current source is supplied to the analog load by means of a repeating series of ON and OFF pulses. The ON-time is the time during which the DC supply is applied to the load, and the OFF-time is the period during which the supply is switched off. Given a sufficient bandwidth, any analog value can be encoded with PWM. At the same time, one of the advantages of PWM is that the signal remains digital all the way from the processor in the controlled system; no digital-to-analog conversion is necessary. By keeping the digital signal, noise effects are minimized.

6.1.2.2 The microcontroller chip

The microcontroller, ATmega16 [243], produced by Atmel Corporation, is a high-performance, low-power eight-bit microcontroller. Fig. 6.4 shows its top view and pin connections. The ATmega16 is a low-power CMOS eight-bit microcontroller based on the AVR-enhanced Reduced Instruction Set Computer (RISC) architecture. By executing powerful instructions in a single clock cycle, the ATmega16 achieves throughputs approaching 1 Million Instructions Per Second (MIPS) per MHz allowing the system designer to optimize power consumption versus processing speed. The ATmega16 has 32 Programmable I/O Lines and 4 PWM channels, so it can perform these three kinds of tasks, which are the acceleration processing, the control instructions calculating, and the PWM signal generating.

Fig. 6.5 is the connection diagram of ATmega16. It can be seen from Figs. 6.3 and 6.5 that: (1) The ATmega16's PB0 and PB1 pins are connected with the MMA7260QT's g-select1 and g-select2 pins, respectively, to set up the sensor's working rage; (2) The ATmega16's PB2 pin is connected to the MMA7260QT's sleepmode pin to set up the sensor's sleep mode; (3) The ATmega16's PA0, PA1, and PA2 pins of are connected with the MMA7260QT's X_{out}, Y_{out}, and Z_{out} pins, respectively, to receive the different direction, acceleration data of the structure collected by the sensor. After the ATmega16 received the acceleration of data structure, it will calculate the control currents of the MR damper according to the control algorithm written in the ATmega16 in advance, and then the decided currents will be outputted to the optical coupler through the ATmega16's PWM pin (the ATmega16's PD4 pin).

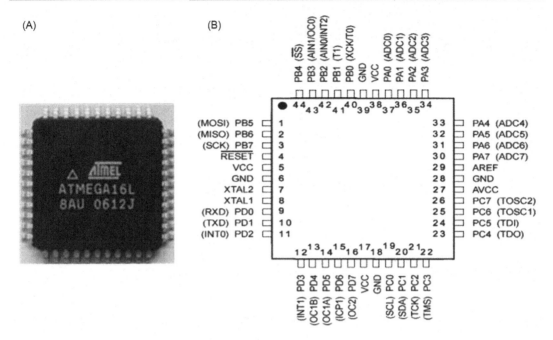

Figure 6.4
ATmega16. (A) Top view and (B) pin connections.

6.1.2.3 The optical coupler

Due to the circuit cannot support the large load voltage, an optical coupler is used to protect the controller from burning out. Optical coupler is a semiconductor device, which is designed to transfer electrical signals by using light waves in order to provide coupling with electrical isolation between circuits or systems. The main purpose of an optical coupler is to prevent rapidly changing voltages or high voltages on one side of a circuit from distorting transmissions or damaging components on the other side of the circuit, and to separate the high-voltage side from the low-voltage side.

In this chapter, the optical coupler, SFH618A, is used. It features a high-current-transfer ratio, low-coupling capacitance, and high-isolation voltage. Fig. 6.6 is the connection diagram of SFH618A.

6.2 Experimental Study on Intelligent Controller

According to the above design principle, the intelligent controller is developed, as shown in Fig. 6.7. In order to grasp and analyze the performance of the intelligent controller, a series of experiments are carried out, including static experiment and dynamic experiment, as shown in Fig. 6.8. In the process of the static experiment, the controller is connected to the

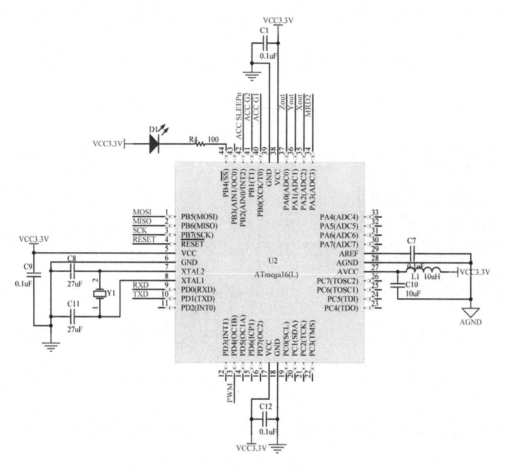

Figure 6.5
The connection diagram of ATmega16.

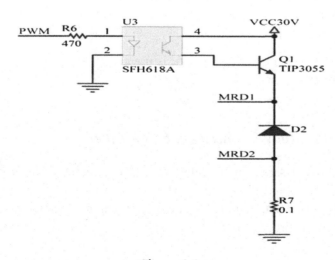

Figure 6.6
The connection diagram of SFH618A.

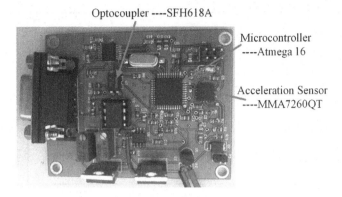

Figure 6.7
The SCM for MR dampers coupling sensing and control.

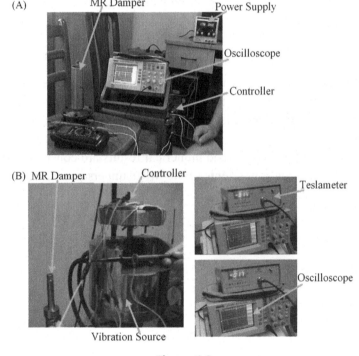

Figure 6.8
The performance experiment of the controller. (A) The static experiment of the controller and (B) the dynamic experiment of the controller.

MR damper, and they are motionless, so the acceleration and the PWM signal are constant. The PWM signal is shown on the oscilloscope, as shown in Fig. 6.8. During the dynamic experiment, the controller is fixed on the top of the actuator of a fatigue testing machine, and it will vibrate with the actuator under the different sine wave excitations. The PWM

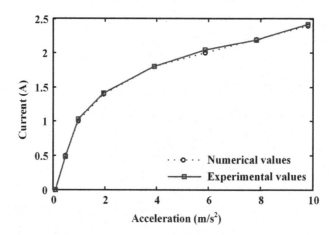

Figure 6.9
The currents comparison between experimental and numerical values.

signal output of the controller and values of the magnetic field intensity produced by the MR damper will be shown on the oscilloscope and the tesla meter in real time, respectively.

Fig. 6.9 shows the comparison between experimental and numerical results of current of the MR damper under different acceleration excitations. It can be seen that the experimental value is 1.032 A, the numerical value is 1.000 A, and the maximum relative error between the experimental and numerical results of current is 3.2% when the acceleration excitation is 0.98 m/s^2. That is, the experimental and numerical results are consistent, and the controller can produce the control currents of the MR dampers in real time in accordance with the measured acceleration data.

CHAPTER 7

Dynamic Response Analysis of the Intelligent Control Structure

Accurate dynamic response analysis of the intelligent control structure is very important for evaluating the control effect of the devices and designs of the devices. Dynamic response analysis involves modeling of the structure and dampers, choosing excitations and analysis method. Generally, dynamic response analysis includes elastic analysis and elasto-plastic analysis.

7.1 Elastic Analysis

For the frame structure incorporated with magnetorheological (MR) dampers shown in Fig. 7.1, MR dampers are connected to the controllers, when the computer receives the state signal measured by the displacement sensor or the acceleration sensor at a certain moment, the computer will issue commands to the MR damper real-time, so as to make the building structure has intelligence and adaptability to the external excitations.

Under the seismic excitation, the equation of motion of the structure incorporated with MR dampers can be expressed as Eq. (2.1). Before calculating the dynamic response of the structure, these matrices in the motion equation should be determined firstly, such as $[M]$, $[C]$, $[K]$, $[B]$, as well as the semiactive control algorithm, these will be discussed as follows.

7.1.1 Mathematical Model of Structures

During the dynamic analysis of the intelligent control structure, the commonly used mathematical model of structures can be classified into two categories, namely the story shear model and the beam-column model.

Multistory building structures or industrial structures are simplified into a story shear model; the simplified multiparticle system is shown in Fig. 7.2. The story shear model is widely used in the structural dynamic analysis and can satisfy the requirement of accuracy for the frame structure due to the large stiffness of the floor.

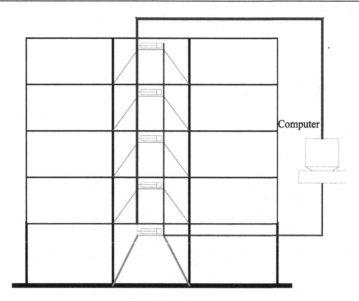

Figure 7.1
Structure incorporated with MR damper.

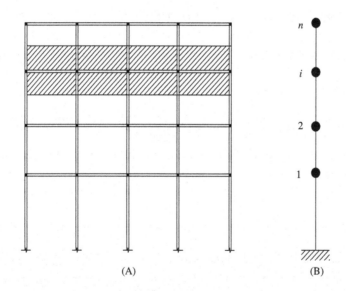

(A)   (B)

Figure 7.2
Story shear model. (A) Multistory building structure (B) Simplified multiparticle system.

The story shear model assumes that the stiffness of the beam and floor of each story is infinite, and the masses of each story are concentrated at the elevation of each floor, the whole lateral stiffness can be calculated by superposition of the antishear members of each floor and only the horizontal displacement of each floor is considered.

The stiffness matrix of the story shear model can be written as:

$$[K] = \begin{bmatrix} K_1 + K_2 & -K_2 & & & \\ -K_2 & K_2 + K_3 & -K_3 & & \\ & & \ddots & & \\ & & & -K_n & K_n \end{bmatrix} \quad (7.1)$$

where K_i is the lateral stiffness of each floor.

The mass matrix of the story shear model can be written as:

$$[M] = \begin{bmatrix} m_1 & & & \\ & m_2 & & \\ & & \ddots & \\ & & & m_n \end{bmatrix} \quad (7.2)$$

where m_i is the mass of each floor.

Compared with the story shear model, the beam-column model is a little more complex, the plane beam-column model is taken for example in this section. When the plane beam model is employed, each node has three degrees of freedom, that is, horizontal displacement u, vertical displacement v, and rotation angle θ, as shown in Fig. 7.3, and it is suitable for the plane frame with strong column and weak beam with small deformation. The axial deformation should be taken to consider the influence of second order effect when large deformation occurs.

The plane bar element is subjected to tension; compression and bending at the same time to the external excitation, however, the influence of the axial deformation on the bending stiffness and the influence of the transverse deformation on the tension and compression stiffness are negligible in the range of small elastic deformation. Thus, the stiffness of the

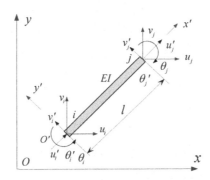

Figure 7.3
Plane bar element.

general bar element can be obtained by superposition of the tension, compression stiffness, and bending stiffness in an appropriate manner:

$$[K_e] = \begin{bmatrix} EA/l & & & & & \\ 0 & 12EI/l^3 & & & & \\ 0 & 6EI/l^2 & 4EI/l & & & \\ -EA/l & 0 & 0 & EA/l & & \\ 0 & -12EI/l^3 & -6EI/l^2 & 0 & 12EI/l^3 & \\ 0 & 6EI/l^2 & 2EI/l & 0 & -6EI/l^2 & 4EI/l \end{bmatrix} \quad (7.3)$$

where A is the cross section area, l is the length, and I is the moment of inertia of the cross section.

The consistent mass matrix and lumped mass matrix are two widely used mass matrices. The former means that the same displacement function is used to establish the stiffness matrix and mass matrix, while the latter means that the distribute mass of the element is lumped at the node. The consistent mass matrix for the plane bar element can be written as follows:

$$[M_e] = \frac{\rho Al}{420} \begin{bmatrix} 140 & & & & & \\ 0 & 156+504R & & & & \\ 0 & (22+42R)l & (4+56R)l^2 & & & \\ 70 & 0 & 0 & 140 & & \\ 0 & 54-504R & (13-42R)l & 0 & 156+504R & \\ 0 & -(13-42R)l & -(3+14R)l^2 & 0 & -(22+42R)l & (4+56R)l^2 \end{bmatrix} \quad (7.4)$$

where ρ is the material density, $R = I/Al^2$ is introduced to consider the influence of rotational inertia.

When the above models are adopted to simulate structure, the damping is usually selected as the orthogonal Rayleigh damping model, the expression is as follows:

$$C = \kappa_1[M] + \kappa_2[K] \quad (7.5)$$

where

$$\kappa_1 = \frac{2\omega_1\omega_2(\zeta_1\omega_2 - \zeta_2\omega_1)}{\omega_2^2 - \omega_1^2} \quad (7.6)$$

$$\kappa_2 = \frac{2(\zeta_2\omega_2 - \zeta_1\omega_1)}{\omega_2^2 - \omega_1^2} \quad (7.7)$$

where ω_1 and ω_2 are the first-order and second-order natural frequencies, respectively; ζ_1 and ζ_2 are the corresponding modal damping ratio and both assumed as 0.05.

7.1.2 Determination of the Control Force of the MR Damper

The classic linear quadratic regular control algorithm has been extensively used for vibration control of structure to determine the optimal control force, which is discussed in detail in Chapter 2, Intelligent Control Strategies. By minimizing the performance index shown in Eq. (2.7), we can obtain the optimal control force.

In Eq. (2.7), Q is the weighting matrix for the structural response and is a $2n \times 2n$ positive-semidefinite matrix; R is the weighting matrix for the control force and is a $m \times m$ positive-definite matrix, n is the number of degree of freedom, m is the number of MR damper. $\{Z(t)\} = \{x(t), \dot{x}(t)\}^T$ is state vector of the controlled system.

The optimal control force vector $\{f_d(t)\}$ for a closed-loop control configuration is written as

$$\{f_d(t)\} = -[R]^{-1}[B]^T[P(t)]\{Z(t)\} = [G]\{Z(t)\} \tag{7.8}$$

where $[G]$ is a feedback gain matrix. Thus, the performance of the LQR algorithm is concerned with the selection of matrices $[Q]$ and $[R]$. In this paper, the weighting matrices $[Q]$ and $[R]$ are selected as [27]

$$[Q] = \alpha \begin{bmatrix} [K] & 0 \\ 0 & [M] \end{bmatrix} \quad [R] = \beta[I] \tag{7.9}$$

where $[I]$ is an $m \times m$ identity matrix; the values of α and β are defined by trial calculation. The diagram of the arrangement of MR dampers is shown in Fig. 7.4.

The desired optimal control forces can be calculated using the sequential quadratic programming algorithm available in MATLAB, and the forces produced by the MR dampers are allocated at the nodes of the structure, the control forces applied on the four nodes can be calculated using the following equation:

$$F_i = F_{i+1} = \frac{F}{2}, \quad F_{i+2} = F_{i+3} = -\frac{F}{2} \tag{7.10}$$

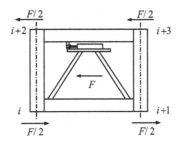

Figure 7.4
Node distribution of damping force.

In order to make the control forces allocate at nodes of structure, $[H]$ should be a $n \times m$ matrix. Element in $[B]$ corresponding to nodes i, $i+1$ and the damper k is given by $[B](i,k) = [B](i+1,k) = 1/2$. Element in $[B]$ corresponding to nodes $i+2$, $i+3$ and the damper k is given by $[B](i+2,k) = [B](i+3,k) = -1/2$. Since the matrix $[B]$ is determined, the damping force matrix $[F]$ of the controlled structure can be represented as follows:

$$[F]_{n \times 1} = [B]_{n \times m} \{f_d(t)\}_{m \times 1} \qquad (7.11)$$

7.1.3 Numerical Analysis

The dynamic responses of the controlled structure cannot be solved by the derivation of equations due to the random characteristics of earthquake excitation. Generally, the dynamic responses, including acceleration, velocity, or displacement responses, can be obtained by the numerical integrating method, i.e., time history analysis method, in which the dynamic response at the time of t_{n+1} can be got based on the dynamic responses at the time of t_n. The commonly used numerical integrating methods include linear acceleration method, *Wilson-θ* method, *Runge-Kutta* method, and so on. In this section, *Wilson-θ* method is chosen for numerical integrating.

The *Wilson-θ* method is an unconditional convergent method, which is proposed by Wilson based on the linear acceleration method. It assumes that the acceleration of the system changes linearly during $\theta \Delta t$. Research shows that this method is unconditionally convergent when $\theta > 1.37$, and θ is always chosen as the 1.4 to prevent from large calculation error.

The detailed calculation processes are shown as follows by taking the calculation step from t_i to t_{i+1}, for example:

1. Calculate the displacement increment $\{\Delta x\}_\tau$ of each particle during a longer period of time $\tau = \theta \Delta t$.
2. Use $\{\Delta x\}_\tau$ to calculate the acceleration increment $\{\Delta \ddot{x}\}$.
3. Calculate the displacement increment $\{\Delta x\}$ and velocity increment $\{\Delta \dot{x}\}$.
4. Compute the relative displacement response $\{x\}_{i+1}$ and velocity response $\{\dot{x}\}_{i+1}$ of t_{i+1} by summation of the response of t_i and the response increment.
5. Substitute $\{x\}_{i+1}$ and $\{\dot{x}\}_{i+1}$ into the equation of motion of the structure to calculate the relative acceleration response of t_{i+1}.

The detailed calculation process can be performed using the following equations:

$$[\overline{K}]_\tau \{\Delta x\}_\tau = \{\Delta \overline{P}\}_\tau \qquad (7.12)$$

$$[\overline{K}]_\tau = [K] + \frac{6}{\tau^2}[M] + \frac{3}{\tau}[C] \qquad (7.13)$$

$$\{\Delta \bar{P}\}_\tau = [M](-\{\Delta \ddot{x}_g\}_\tau + \frac{6}{\tau}\{\dot{x}\}_i + 3\{\ddot{x}\}_i) + [C](3\{\dot{x}\}_i + \frac{1}{2}\{\ddot{x}\}_i \tau) \tag{7.14}$$

$$\{\Delta \ddot{x}\} = \frac{1}{\theta}\{\Delta \ddot{x}\}_\tau = \frac{6}{\theta \tau^2}\left[\{\Delta x\}_\tau - \{\dot{x}\}_i \tau - \frac{1}{2}\{\ddot{x}\}_i \tau^2\right] \tag{7.15}$$

$$\{\Delta \dot{x}\} = \{\ddot{x}\}_i \Delta t + \frac{1}{2}\{\Delta \ddot{x}\}\Delta t \tag{7.16}$$

$$\{\Delta x\} = \{\dot{x}\}_i \Delta t + \frac{1}{2}\{\ddot{x}\}_i \Delta t^2 + \frac{1}{6}\{\Delta \ddot{x}\}\Delta t^2 \tag{7.17}$$

$$\{x\}_{i+1} = \{x\}_i + \{\Delta x\} \tag{7.18}$$

$$\{\dot{x}\}_{i+1} = \{\dot{x}\}_i + \{\Delta \dot{x}\} \tag{7.19}$$

$$\{\ddot{x}\}_{i+1} = -([M]^{-1}[C]_{i+1}\{\dot{x}\}_{i+1} + [M]^{-1}[K]_{i+1}\{x\}_{i+1} + \{\ddot{x}_g\}_{i+1}) \tag{7.20}$$

7.2 Elasto-Plastic Analysis Method

Plastic deformation at the end of the member will be produced under strong wind and earthquake excitations, which will change the stiffness and damping of the structure. Therefore, elasto-plastic time-history analysis should be employed to solve this problem. The calculation process of the elasto-plastic analysis is similar to the elastic analysis; however, the restoring force model and elasto-plastic stiffness matrix should be determined due to the reason that they vary with time. In view of this, this section will focus on the introduction of the restoring force model and the determination of elasto-plastic stiffness matrix.

7.2.1 Restoring Force Model

The structure usually exhibits nonlinear characteristics in elasto-plastic analysis, i.e., stiffness and damping of the structure are parameters changing with time. Thus, the restoring force model should be employed to describe the restoring force characteristics. The widely used models are double linear model and threefold line stiffness retrograde model. The double linear model is relatively simple and is suitable for simulating the steel structure. In order to clearly reflect the working state of the elasto-plastic characteristics of the reinforced concrete structure, the threefold line stiffness retrograde model [244] is employed in this section, the corresponding restoring force characteristic curve is shown in Fig. 7.5.

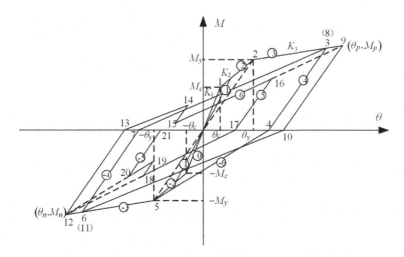

Figure 7.5
Threefold line stiffness retrograde model.

As shown in Fig. 7.5, point 1 is cracking point, the corresponding force and deformation are M_c and θ_c; points 2 and 5 are yielding points, the corresponding force and deformation are M_y and θ_y, the unloading lines 3–4 and 9–10 after yield are parallel to secant 0–2, k_4 is the secant stiffness. After a cyclic loading, the following conclusions can be obtained: (1) the decrement of stiffness is relevant to the maximum deformation of the previous cyclic loading; (2) the straight line of the reverse loading directs to the maximum deformation point of the previous deformation; and (3) the unloading stiffness after yield is equal to the secant stiffness.

It can be also seen that the member is at point 7 after a cyclic loading, when loading again, point 7 will direct to the maximum deformation point 3, the unloading process is along the lines 9–10, then direct to point 6 when reverse loading at point 10. Usually, the threefold line stiffness retrograde model can be determined using five parameters, namely, cracking force M_c, yield force M_y, elastic stiffness k_1, cracking stiffness k_2, and yield stiffness k_3.

In order to programming the threefold line stiffness retrograde model, the hysteretic curve is divided into 12 parts. Then the corresponding transition conditions and logical relations are shown in Fig. 7.6, and the turning points need specific processing.

7.2.2 Processing of Turning Points

The state transition manifests at the intersection of two straight lines of the restoring force model, such as points 1, 2, 3, 4, 9, and 10 shown in Fig. 7.5. Point E shown in Fig. 7.7A is taken for example, assuming that the structure is located at point D at time t, then arrives at G in next time $t + \Delta t$; however, it deviates from the restoring force curves obviously.

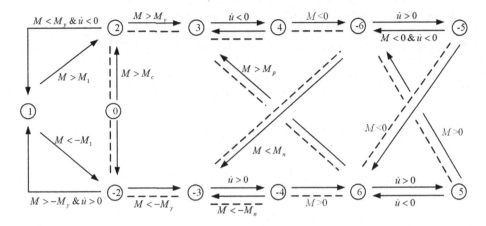

Figure 7.6
The transition relation for the elasto-plastic analysis of the beam-column model.

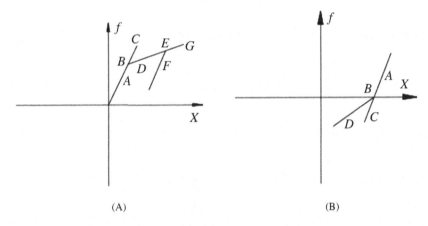

Figure 7.7
Processing of the turning points. (A) The first and second kind and (B) the third kind.

Actually, the structure is located at point G at time $t + \Delta t$. The same situation exists at point B. Thus, the steady error accumulation will make the restoring force model distortion and lead to wrong results if the turning points are not properly processed. In view of this, a reduced time step is employed to tackle this problem, that is $\Delta t' = p\Delta t$ $(0 < p < 1)$.

Based on the above discussion, the most important is to determine the value of p in processing the turning points, which can be classified as three categories according to the processing method. The first kind is the transition points from steep to slow, such as points 1, 2, 5, and 16. The second kind is the transition points from slow to deep, such as points 3, 6, 8, 9, 12, and 14. The third kind is the change of the restoring force direction, such as points 4, 7, 10, 13, 17, and 21.

7.2.2.1 Determination of p for the first kind of turning point

As shown in Fig. 7.7A, by using Taylor series expansion, the displacement of point B can be written as:

$$x_B = x(t + \Delta t) = x(t) + \dot{x}(t)p\Delta t \tag{7.21}$$

Then, p can be determined using the following equation:

$$p = \frac{x(t + p\Delta t) - x(t)}{\dot{x}(t)\Delta t} = \frac{x_B - x_A}{\dot{x}_A \Delta t} = \frac{x_B - x_A}{x_C - x_A} \tag{7.22}$$

Finally, the velocity and acceleration of point B can be written as:

$$\dot{x}_B = \dot{x}_A + p(\dot{x}_c - \dot{x}_A) \tag{7.23}$$

$$\ddot{x}_B = \ddot{x}_A + p(\ddot{x}_c - \ddot{x}_A) \tag{7.24}$$

7.2.2.2 Determination of p for the second kind of turning point

As shown in Fig. 7.7A, taking point E, for example:

$$\dot{x}_E = \dot{x}_{t+P\Delta t} = \dot{x}(t) + \Delta\dot{x}(P\Delta t) = 0 \tag{7.25}$$

$$\Delta\dot{x}(P\Delta t) = -\dot{x}(t) = \ddot{x}(t)P\Delta t + \frac{\Delta\ddot{x}}{\Delta t} \cdot \frac{(P\Delta t)^2}{2} \tag{7.26}$$

7.2.2.3 Determination of p of the third kind of turning point

As shown in Fig. 7.7B, taking point B, for example:

$$P = \frac{f_c}{f_c - f_A} \tag{7.27}$$

7.2.3 Elasto-Plastic Stiffness Matrix

The computational models of the elasto-plastic element of the beam-column model can be mainly classified as: single component model, double component model, and three-component model. The single component member system model is employed as the computational model in this section. As shown in Fig. 7.8, the inflection point is supposed to be the member midpoint and vertical offset is supposed to be negligible for member ends bending moment. According to Giberson single component principle [245], the ith member end bending moment increment ΔM_i is given by

$$\Delta M_i = S_{ij}(\Delta\theta_i - \Delta\alpha_i) + S_{ji}C_{ji}(\Delta\theta_j - \Delta\alpha_j) \tag{7.28}$$

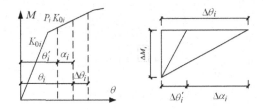

Figure 7.8
Skeleton curve of hysteresis model at the end of member.

where S_{ij} and S_{ji} are flexural rigidity, which takes into account the shear deformation. $S_{ij} = S_{ji}(2EI(2+\gamma))/(L(1+2\gamma))$, $\gamma = 6\mu EI/GAL^2$, $C_{ij} = (1-\gamma)/(2+\gamma)$ with regard to the prismatic member, $\Delta\theta_i$ and $\Delta\theta_j$ are the ith and jth end angle increment, $\Delta\alpha_i$ and $\Delta\alpha_j$ are the ith and jth end plastic angle increment. Member dual-end elastic angle increment can be expressed as $\Delta\theta_i - \Delta\alpha_i = \Delta\theta_j - \Delta\alpha_j$, since the inflection point is supposed to be the midpoint. Thus, Eq. (7.28) can be written as

$$\Delta M_i = K_{0i}(\Delta\theta_i - \Delta\alpha_i) \tag{7.29}$$

where $K_{0i} = S_{ij}(1 + C_{ij})$ is the equivalent flexural rigidity.

As shown in Fig. 7.8, plastic angle increment can be obtained as

$$\Delta\alpha_i = \frac{\Delta M_i}{P_i K_{0i}} - \frac{\Delta M_i}{K_{0i}} = \left(\frac{1-P_i}{P_i}\right)\frac{\Delta M_i}{K_{0i}} \tag{7.30}$$

Substituting Eq. (7.30) into (7.29), we can obtain the following equation:

$$\Delta M_i = P_i K_{0i} \Delta\theta_i \tag{7.31}$$

When taking plane deformation into account exclusively [246], the relationship between force increment and the displacement increment of the elasto-plastic member end can be written as

$$\{\Delta F\}^e = [k]^e_{ep}\{\Delta\delta\}^e \tag{7.32}$$

where $\{\Delta\delta\}^e$ is member end total displacement increment and $[k]^e_{ep}$ is unit elasto-plastic stiffness matrix.

$$[k]^e_{ep} = \begin{bmatrix} e & 0 & 0 & -e & 0 & 0 \\ 0 & a & b_i & 0 & -a & b_j \\ 0 & b_i & c_i & 0 & -b_i & d \\ -e & 0 & 0 & e & 0 & 0 \\ 0 & -a & -b_i & 0 & a & -b_j \\ 0 & b_j & d & 0 & -b_j & c_j \end{bmatrix} \tag{7.33}$$

$$e=\frac{EA}{L}, \quad a=\frac{6EI}{L^3}\times\frac{P_i+P_j}{\beta}, \quad b_i=\frac{6EI}{L^2}\times\frac{P_i}{\beta}, \quad b_j=\frac{6EI}{L^2}\times\frac{P_j}{\beta}, \quad c_i=\frac{4EI}{L}\times\frac{P_i[3-(1-\gamma)P_j]}{\beta},$$

$$c_j=\frac{4EI}{L}\times\frac{P_j[3-(1-\gamma)P_i]}{\beta}, \quad d=\frac{2EI}{L}\times\frac{(1-\gamma)P_iP_j}{\beta}, \quad \beta=3-(1-\gamma)(P_i+P_j)$$

where P_i and P_j are the stiffness reduction coefficients of the two ends of each member, and the value of P_i and P_j is changing in the time history analysis procedure, they are adopted to change the stiffness of the member.

Based on the above theoretical discussion, the dynamic response analysis of intelligent-controlled structure can be carried out using the time-history method, and specific numerical analysis examples will be introduced in Chapter 8, Example and Program Analysis.

7.3 Dynamic Response Analysis by SIMULINK

7.3.1 Simulation of the Controlled Structure

In Chapter 2, Intelligent Control Strategies, the equation of motion of the civil structure with energy dissipation devices under dynamic loads has been given as Eq. (2.5)

$$\begin{cases} \{\dot{Z}(t)\} = [A]\{Z(t)\} + [D_0]\{P(t)\} + [B_0]\{f_d(t)\} \\ \{Y\} = [E_0]\{Z(t)\} \end{cases} \quad (7.34)$$

It is simple to use the SIMULINK toolbox of MATLAB to calculate the dynamic responses of the structure.

This section takes the controlled structure with MR damper for example, and focuses on the methods to deal with such problems using SIMULINK toolbox. SIMULINK is a software package of MATLAB, which is used for modeling, simulating, and analyzing the dynamic system. It is intuitive, simple, and easy to understand.

In this section, the control force $\{f_d(t)\}$ provided by MR dampers is calculated by LQR control strategy, which is introduced in Chapter 2, Intelligent Control Strategies. According to Eq. (2.5), the state-space equation of the structure with MR dampers is given as

$$\begin{cases} \{\dot{Z}(t)\} = [A]\{Z(t)\} - [D_0]\{\Gamma\}\ddot{x}_g(t) + [B_0]\{f_d(t)\} \\ \{Y\} = [E_0]\{Z(t)\} \end{cases} \quad (7.35)$$

where $\{\Gamma\}$ is a column vector of ones, $\ddot{x}_g(t)$ is the earthquake acceleration excitation, the $\{f_d(t)\}$ can be got as Eq. (2.17):

$$\{f_d(t)\} = -[G]\{Z(t)\} \quad (7.36)$$

where $\{G\} = -[R]^{-1}[B_0]^T\{\tilde{P}(t)\}$. It can be got by LQR function in MATLAB.

$$[G] = \text{LQR}([A], [B_0], [Q], [R]) \tag{7.37}$$

Eq. (7.35) can be written as

$$\begin{cases} \{\dot{Z}(t)\} = ([A] - [B_0][G])\{Z(t)\} - [D_0]\{\Gamma\}\ddot{x}_g(t) \\ \{Y\} = [E_0]\{Z(t)\} \end{cases} \tag{7.38}$$

where $[A] = \begin{bmatrix} [0] & [I] \\ -[M]^{-1}[K] & -[M]^{-1}[C] \end{bmatrix}$ is the system matrix, $[B_0] = \begin{bmatrix} [0] \\ -M^{-1}B \end{bmatrix}$ is the input matrix, $[Q] = \alpha \begin{bmatrix} K & 0 \\ 0 & M \end{bmatrix}$, $[R] = \beta[I]$, $[0]$ is the null matrix, $[I]$ is the identity matrix, α, β are the weighting factor.

Based on the state-space system described above, the SIMULINK model of the controlled structure can be drawn as shown in Fig. 7.9.

7.3.2 Numerical Analysis

In this example, a three-floor reinforced concrete frame structure with dampers is modeled and simulated by SIMULINK. The mass of every floor of the building is 5×10^5 kg and the shear stiffness is 3×10^8 N/m. The MR dampers are installed on each floor of the structure, as shown in Fig. 7.10. El Centro earthquake with 200 gal acceleration amplitude is selected, and the sampling time is 0.02 s.

For the three-story structure equipped with dampers shown in Fig. 7.10, SIMULINK toolbox in MATLAB can be used to get dynamic responses of the controlled structure. Specific steps are as follows:

1. Write the equations of motion of the controlled structure into the form of state-space equations, such as Eqs. (7.36) and (7.38).

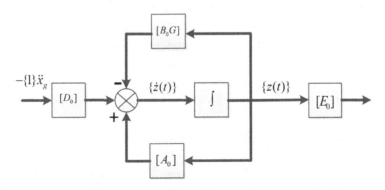

Figure 7.9
The SIMULINK model of the controlled structure.

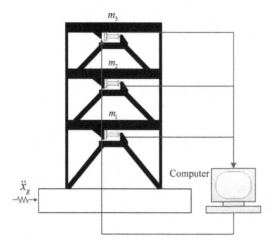

Figure 7.10
Schematic of the intelligent control structure.

2. Start MATLAB and SIMULINK toolbox, and then open the SIMULINK Library Browser window and SIMULINK Model Editor, as shown in Fig. 7.11.
3. According to state-space equations (7.36) and (7.38), select and establish the modules required in the SIMULINK Library Browser window. Select ▸⟦⅟ₛ⟧▸—*Integrator module*— in the *Continuous* under *SIMULINK* directory, select ▸▷▸—*Gain module* and ⟦+⟧—*Add module* in the *Math Operations* under *SIMULINK* directory, select ⟦simin⟧▸—*From Workspace module* in the *Sources* and ▸⟦□⟧—*Scope module* in the *Sinks* under the *SIMULINK* directory, and drag the selected modules into SIMULINK model editor using the mouse, as shown in Fig. 7.12.
4. According to state-space equations (7.36) and (7.38), the SIMULINK model of the controlled structure is shown in Fig. 7.9. Establish SIMULINK model of the controlled structure in the SIMULINK model editor, as shown in Fig. 7.13.
5. Determine SIMULINK model parameters of each module of the structure and modify the name of each module. As each module parameter of the example is high-dimension matrix, it is very inconvenient to add or modify parameters directly in the module dialog. Therefore, it is convenient to run M-files or DAT data files to output the parameters required for each module to the workspace, and rename the module in the parameter dialog.

Attention: To facilitate viewing and debugging program, the module can be renamed as shown in Fig. 7.14.

As shown in Fig. 7.14, "dzb" block is an earthquake excitation source, which is used to input the seismic wave to the structure from MATLAB workspace that is $\{\ddot{x}_g(t)\}$ as

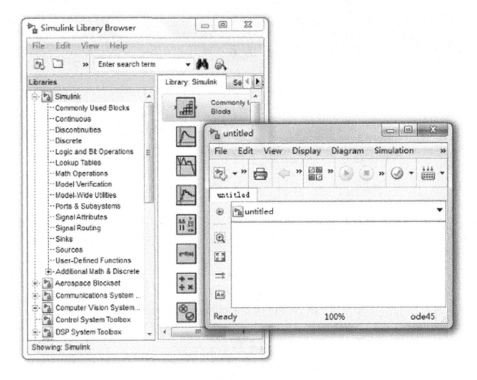

Figure 7.11
SIMULINK Library Browser window and SIMULINK Model Editor.

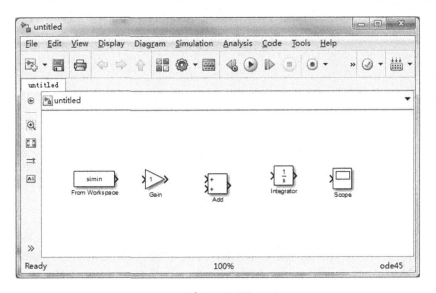

Figure 7.12
Selected modules.

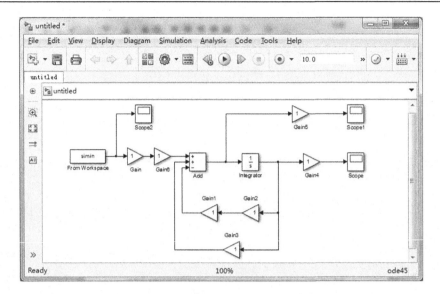

Figure 7.13
SIMULINK model of the controlled structure.

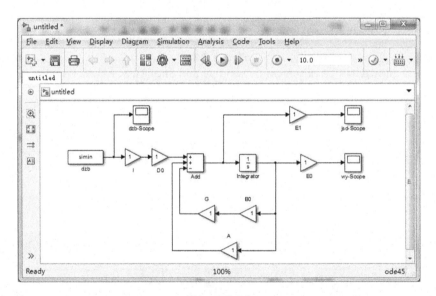

Figure 7.14
Module renamed of the SIMULINK model diagram system.

shown in Eq. (7.38). Double click the "dzb" block using the mouse to open the dialog box and modify its default value to "dzb", which is shown in Fig. 7.15.

The "dzb-Scope" block is used to display the seismic wave; the parameter of "I" block corresponds to $\{\Gamma\}$ in Eq. (7.38) and modifies its default value to "I" in the

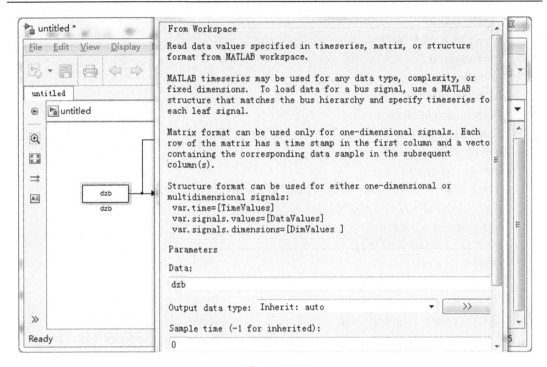

Figure 7.15
Setting the "dzb" block.

dialog box; the parameter of "D0" block corresponds to [D_0] in Eq. (7.38) and modifies its default value to "D0" in the dialog box; the parameter of "G" block corresponds to [G] in Eq. (7.38) and modifies its default value to "G" in the dialog box; the parameter of "B0" block corresponds to [B_0] in Eq. (7.38) and modifies its default value to "B0" in the dialog box; the parameter of "A" block corresponds to [A] in Eq. (7.38) and modifies its default value to "A" in the dialog box; the "E0" block and "E1" block are the matrices to be used to extract the displacement and acceleration responses of the third floor of the structure; the "jsd-Scope" block and "wy-Scope" block are used to display the displacement and acceleration response of the third floor of the structure.

The following is the main program to form parameters of the module:
dzb
% generating seismic waves. The parameters of "dzb" block is generated.
% enter the parameters of the structure.
m0 = [5 5 5]*1e + 5;
k0 = [3 3 3]*1e + 8;
zxznb1 = 0.05;

```
zxznb2 = 0.07;
c0 = [35.1,35.1,35.1];
c1 = diag(c0);
m = diag(m0);
cn = length(m0);
[k] = matrixju(k0,cn);
[x,d] = eig(k,m);
d = sqrt(d);
w = sort(diag(d));
a = 2*w(1)*w(2)*(zxznb1*w(2)-zxznb2*w(1))/(w(2)^2-w(1)^2);
b = 2*(zxznb2*w(2)-zxznb1*w(1))/(w(2)^2-w(1)^2);
c = a*m + b*k + c1;
Bs = [1 -1 0;0 1 -1;0 0 1];
[n n] = size(m);
A = [zeros(n) eye(n);-inv(m)*k -inv(m)*c1];      % the parameters of A block
B0 = [zeros(n); inv(m)*Bs];                      % the parameters of B0 block
D0 = [zeros(n,1);-ones(n,1)];                    % the parameters of D0 block
I = ones(1);                                     % the parameters of I block
Q = 100*[k zeros(n);zeros(n) m];
R = 8e-6*eye(n);
G = lqr(A,B0,Q,R);                               % the parameters of G block
E0 = [0 0 1 0 0 0];                              % the parameters of E0 block
E1 = [0 0 0 0 0 1];                              % the parameters of E1 block
```

The **matrixju.m** is called subroutine in this main program which is listed below.

```
function[kcju] = matrixju(korc,cn);
kcju = zeros(cn);
for i = 1:cn-1;
        kcju(i,i) = korc(i) + korc(i + 1);
        kcju(i,i + 1) = -korc(i + 1);
        kcju(i + 1,i) = -korc(i + 1);
end
kcju(cn,cn) = korc(cn);
```

6. Modifying the simulation parameters. Using mouse to click the "model configuration parameters" on the SIMULINK model editor menu bar to open the simulation parameters dialog, and modify the simulation end time to "10", which represents that the simulation time is set to 10 s as shown in Fig. 7.16.

7. The simulation of the model. First, we need to run the above main program to get the parameters required in the SIMULINK model. Then use the mouse to click the

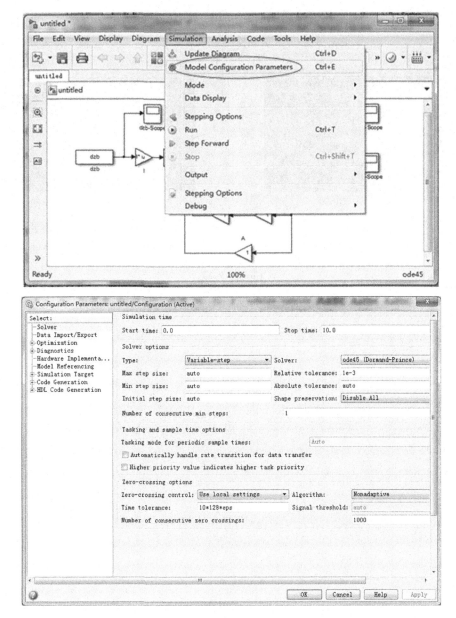

Figure 7.16
Modify the simulation parameters.

button ▶ on the *SIMULINK Model Editor* menu bar or click *Start* command on the *Simulation* menu to run the simulation. Finally, use the mouse to click the "jsd-Scope" block and "wy-Scope" block, the displacement and acceleration responses of the structure can be displayed as shown in Fig. 7.17.

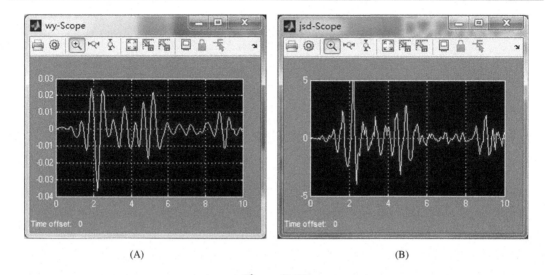

Figure 7.17
The results of SIMULINK simulation. (A) The displacement responses of the structure and (B) the acceleration responses of the structure.

CHAPTER 8

Example and Program Analysis

How to solve the real problems of intelligent vibration control in civil engineering structures is a comprehensive application of knowledge described in the previous chapters. Some examples about intelligent vibration control in different civil engineering structures with different kinds of intelligent control devices will be presented to narrate semiactive control and active control.

8.1 Dynamic Analysis on Frame Structure With MR Dampers

8.1.1 Structural and Damper Parameters

A 10-story reinforced concrete frame structure in the 8 degree area is chosen to evaluate the control effect of the magnetorheological (MR) damper. The height of the bottom layer is 5 m, the height of the other layers is all 3.2 m, and the spans from left to right are 8.4 m, 5.4 m, and 7.2 m, respectively. The cross section of the column for the first to the sixth story is 600×600 mm, and 500×500 mm for 7–10 layers. The sizes of the beams are 250×550 mm, 250×600 mm, and 250×650 mm, respectively, and the damping ratio is chosen as 0.05. The structure diagram and the reinforcement figures are shown in Fig. 8.1.

The elastic time-history analysis is performed using the beam-column model. The MR dampers are installed in the middle span in the bottom six floors, as shown in Fig. 8.2. El Centro wave and Taft wave with a modified peak value of 140 gal are chosen as earthquake excitations. The maximum force of the MR damper is 200 KN, and the viscous damping coefficient is $c_d = 60$ KN s/m. The consistent mass matrix and Rayleigh damping matrix are employed. The dynamic responses of the structure under the condition of passive-off control, semiactive control, and uncontrolled are calculated by MATLAB programming, among which the passive-off control means the MR damper control with no current input and only the viscous damping force is provided.

8.1.2 Semiactive Control Strategy

MR damper, as a semiactive control device, cannot provide the optimal control force at any instant, thus the following control strategy is adopted to realize semiactive control, as shown in Eq. (8.1). It can be seen that the damper will provide the maximum force when the optimal control force is larger than the maximum force and its direction is different

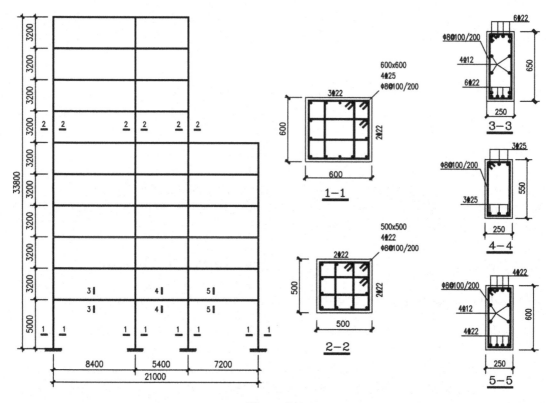

Figure 8.1
Structure calculation diagram.

from the velocity of the damper; similarly, the damper will provide the minimum force when the optimal control force is smaller than the minimum force and its direction is the same as the velocity of the damper; for other conditions, the damper will provide the optimal control force, which is calculated using the linear quadratic regulator (LQR) optimal control algorithm discussed in Chapter 2, Intelligent Control Strategies.

$$F_i = \begin{cases} \text{sgn}(u_i)F_{i,\max} & F_{i,\text{opt}} \cdot \dot{u}_i < 0 \text{ and } |F_{i,\text{opt}}| > F_{i,\max} \\ F_{i,\text{opt}} & F_{i,\text{opt}} \cdot \dot{u}_i < 0 \text{ and } F_{i,\min} < |F_{i,\text{opt}}| < F_{i,\max} \\ \text{sgn}(u_i)F_{i,\min} & |F_{i,\text{opt}}| < F_{i,\min} \text{ or } F_{i,\text{opt}} \cdot \dot{u}_i > 0 \end{cases} \quad (8.1)$$

where $F_{i,\min}$, $F_{i,\max}$ are the minimum and maximum damping force provided by the ith MR damper, u_i, $\dot{u}_i$ are the displacement and velocity of the ith MR damper, $F_{i,\text{opt}}$ is the optimal control force calculated by LQR optimal control algorithm, F_i is the damping force provided by the ith MR damper. Fig. 8.3 shows the semiactive control strategy.

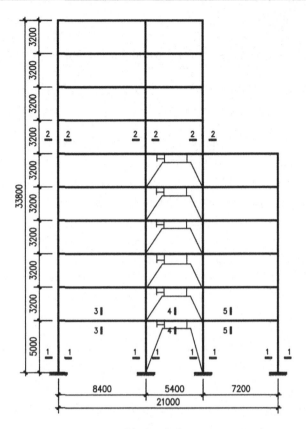

Figure 8.2
Installation of MR dampers.

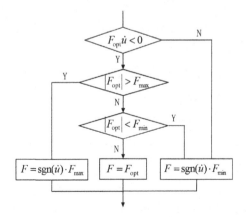

Figure 8.3
Semiactive control strategy.

8.1.3 Results and Analysis

Dynamic responses of the structures with and without MR dampers can be got through elastic time-history analysis. Fig. 8.4 shows the displacement and acceleration time-history responses of the top node (node 40). It can be seen that the peak values of the displacement responses are reduced by 23.4% and 44.1%, respectively, for the passive-off control and the semiactive control compared with the uncontrolled structure, which shows the semiactive control strategy is more effective to control the displacement response. For acceleration response, the semiactive control effect is better than the passive-off control effect in the view of the whole time-history curves. While the peak values of acceleration responses are reduced by 15.3% and 12.2%, respectively, for the passive-off control and semiactive control, this is mainly because the semiactive control strategy used in this example is based on the displacement and velocity response feedback method, and the increase of control current of MR dampers will increase the stiffness of the structure, this will lead to increase of acceleration response in some earthquake excitations. Totally, the semiactive control can significantly reduce the displacement responses and has a much better control effect than the passive-off control.

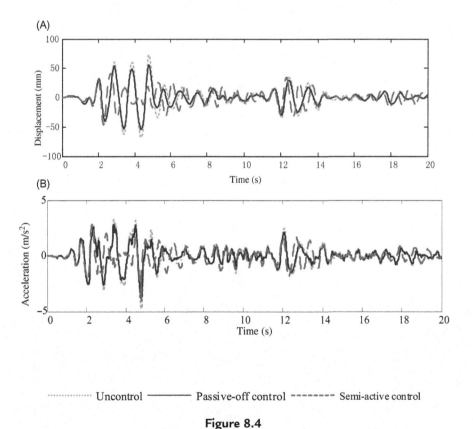

Figure 8.4

Dynamic response of the node 40 under El Centro wave. (A) Displacement under El Centro wave and (B) Acceleration under El Centro wave.

The displacement and acceleration envelop diagram of each floor under El Centro and Taft waves are also calculated, as shown in Figs. 8.5 and 8.6, respectively. It can be seen that the semiactive control strategy has a good control effect compared with the passive-off control. The peak displacement responses of the semiactive controlled structure is far smaller than those of the passive-off controlled and uncontrolled structures, which means that the current is adjusted constantly according to the dynamic responses of the structure to make the damping forces of MR dampers close to the optimal control forces. Similarly, the acceleration response of the semiactive controlled structure is smaller than the passive-off controlled structure in general, except for the top three stories.

The time-history curves of the control force of the MR damper under the El Centro and Taft waves are shown in Fig. 8.7, and this shows that the control force of the MR damper will change real-time during the earthquake excitation.

In order to evaluate the control effect of semiactive control and passive control quantitatively, the following indicators are adopted to evaluate the control effect, as listed in Table 8.1.

The results of the two indicators of the second, fourth, sixth, eighth, and tenth stories under different seismic excitations are listed in Table 8.2. It can be seen that the displacement and

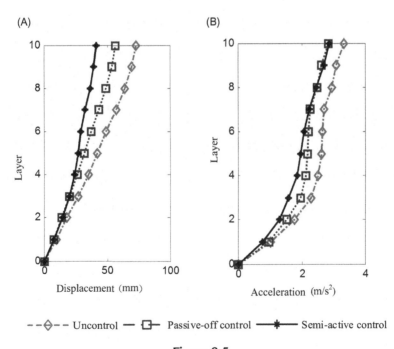

Figure 8.5
Envelop of the (A) displacement and (B) acceleration responses of the layers under El Centro wave.

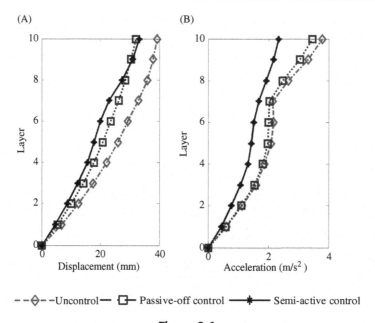

Figure 8.6
Envelop of the (A) displacement and (B) acceleration responses of the layers under Taft wave.

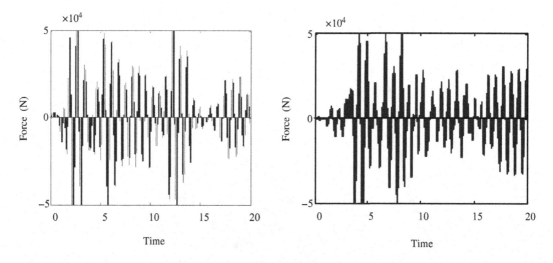

Figure 8.7
Time-history curves of the control force of the MR damper in the first layer. (A) El Centro wave and (B) Taft wave.

Table 8.1: Evaluation indicators of the control effect

Category	Formula	Meaning
Peak response indicators	$J_1 = \max\left\{\dfrac{|x_{i0}| - |x_{ic}|}{|x_{i0}|}\right\}$	Displacement peak value
	$J_2 = \max\left\{\dfrac{|\ddot{x}_{i0}| - |\ddot{x}_{ic}|}{|\ddot{x}_{i0}|}\right\}$	Absolute acceleration peak value

where x_{i0} and x_{ic} are the displacement responses of the ith story of the uncontrolled and controlled structures; $\ddot{x}_{i0}$ and $\ddot{x}_{ic}$ are the acceleration responses of the ith story of the uncontrolled and controlled structures.

Table 8.2: Indicator value under two excitations

Indicator	Control Method	El Centro Wave					Taft Wave				
		2	4	6	8	10	2	4	6	8	10
J_1(%)	Passive-off	23.1	25.2	24.0	22.9	22.7	19.3	19.4	19.7	20.5	17.8
	Semiactive	24.8	31.3	43.3	43.1	43.1	29.7	29.9	31.3	23.1	13.9
J_2(%)	Passive-off	14.4	15.4	16.6	14.6	15.1	2.73	3.37	7.1	7.04	8.78
	Semiactive	26.5	31.7	17.7	11.9	13.7	31.1	29.3	30.7	27.7	38.7

acceleration responses are reduced effectively by using both semiactive and passive-off control method. Specifically, under El Centro earthquake excitation, the displacement responses of the semiactive controlled structure is reduced by 1.7%, 6.3%, 19.3%, 20.2%, and 20.6%, respectively, compared with those of the passive-off controlled structure. The acceleration responses of the second, fourth, and sixth stories of the semiactive controlled structure is reduced more effectively than those of the passive-off controlled structure, while the responses of the passive-off controlled structure are reduced more effectively at the eighth and tenth stories with a degree of 2.7% and 1.4%, respectively.

Similarly, the displacement responses of the semiactive controlled structure are also reduced effectively compared with the passive-off controlled structure under Taft earthquake excitation, except for the tenth story. In addition, the reduction rates of acceleration responses of the semiactive controlled structure are larger than the passive-off controlled structure by 29.4%, 26%, 23.6%, 20%, and 30%, respectively.

This section aims at the vibration control of building structure using the MR damper; the dynamic responses of the uncontrolled, passive-off controlled, and semiactive controlled structures are calculated. Comparison results show that dynamic responses of the semiactive controlled structure is smaller than those of the passive-off controlled and uncontrolled structure, and MR dampers can effectively dissipate the vibration energy and reduce the dynamic responses of the building structure.

8.2 Dynamic Analysis on Long-Span Structure With MR Dampers

The cable-stayed bridges are easy to vibrate under various excitations due to the large flexibility, small quality, and small damping. In the latest decade, MR dampers are widely used to mitigate the vibration of cable-stayed bridges. For further study of wind vibration control of cable-stayed bridges using the MR damper, the plane truss model of a cable-stayed bridge is established in this section, and the dynamic responses of the controlled and uncontrolled bridges are calculated.

8.2.1 Parameters and Modeling

The span of the cable-stayed bridge is 50 m + 125 m + 50 m, and the height of the main tower is 30 m. The following assumptions are adopted to establish the plane truss model of the cable-stayed bridge.

1. The bridge tower is consolidated with the bridge pier.
2. The degrees of freedom in the vertical and rotation direction at the joint of the bridge girder and the tower are coupled, and the degree of freedom along the bridge is released.
3. The bridge girder is hinged supported at the two ends.
4. The cables are hinged at the girder and tower.

The angles of the cables range from 26.5 to 72 degrees, the basic parameters of the cable-stayed bridge are listed in Table 8.3, and the model is shown in Fig. 8.8.

In order to verify the accuracy of the model established by MATLAB, the natural frequencies are compared with the finite element software ANSYS. The finite element model is shown in Fig. 8.9, Beam3 element is employed to model the bridge girder and the main tower; the Link10 element is used to model the cables; the initial prestress is simulated through the initial strain in Link1 element.

The first ten natural frequencies of the cable-stayed bridge are listed in Table 8.4, and comparison results show that the MATLAB results agree well with the ANSYS results, which confirms the fine precision of the plane truss model established by MATLAB. In view of this, this MATLAB model can be used to perform the wind vibration analysis of the cable-stayed bridge incorporated with MR dampers.

Table 8.3: Basic parameters of the cable-stayed bridge

	Element	Section Area A (m^2)	Elastic Modulus E (N/m^2)	Inertia Moment I (m^4)	Density ρ (kg/m^3)
Girder	Beam	0.12	2.1×10^{11}	0.016	2.7×10^3
Tower	Beam	0.08	2.1×10^{11}	0.013	3.2×10^3
Cable	Bar	0.006	1.6×10^{11}	—	0.5×10^3

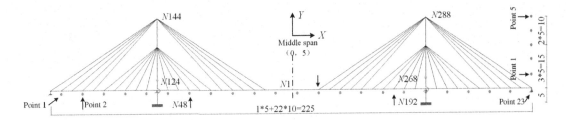

Figure 8.8
Location of the wind velocity simulation points (unit: m).

8.2.2 Wind Load Simulation

In order to perform the wind vibration analysis, the wind speed time history should be obtained firstly, the harmonic superposition method [247] is adopted to simulate the wind velocity history. The calculation processes are as follows:

1. Taking the Cholesky decomposition of $[S^0(\omega)]$

$$[S^0(\omega)] = [H(\omega)][H(\omega)]^{T*} \tag{8.2}$$

where $[S^0(\omega)]$ is the spectrum density matrix; $[H(\omega)]$ is a lower triangle matrix; $[H(\omega)]^{T*}$ is the conjugate transpose matrix of $[H(\omega)]$, and it can be expressed as follows:

$$[H(\omega)] = \begin{bmatrix} H_{11}(\omega) & 0 & \cdots & 0 \\ H_{21}(\omega) & H_{22}(\omega) & \cdots & 0 \\ \cdots & \cdots & \cdots & \cdots \\ H_{n1}(\omega) & H_{n2}(\omega) & \cdots & H_{nn}(\omega) \end{bmatrix} \tag{8.3}$$

2. Solving the random phase with related characteristic, the random phase can be expressed as,

$$\theta_{jm}(\omega) = \tan^{-1}\left\{\frac{\text{Im}[H_{jm}(\omega)]}{\text{Re}[H_{jm}(\omega)]}\right\} \tag{8.4}$$

3. The wind speed time history of the simulated points can be calculated using the following equation according to the Shinozuka theory.

$$u_i(t) = \sum_{j=1}^{i}\sum_{k=1}^{N} |H_{ij}(\omega_k)|\sqrt{2\Delta\omega}\cos[\omega_k t - \theta_{ij}(\omega_k) + \varphi_{jk}] \tag{8.5}$$

where N is the sampling points, $\Delta\omega$ is the frequency spacing, φ_{jk} is the phase angle among $0 \sim 2\pi$ with uniform distribution.

210 Chapter 8

(A)
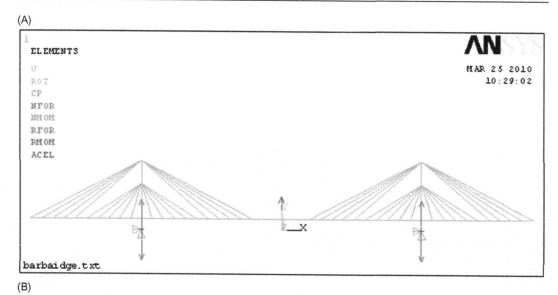

(B)
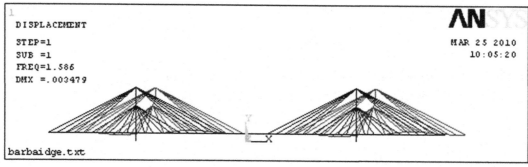

(C)
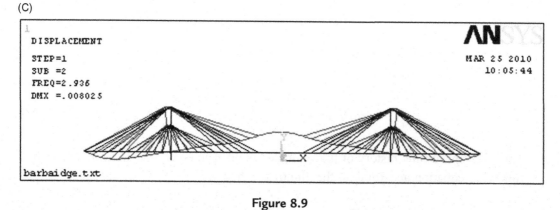

Figure 8.9

The ANSYS model and modal analysis results. (A) Diagram of the cable-stayed bridge, (B) the first mode shape, and (C) the second mode shape.

Table 8.4: Comparison of the natural frequencies

Mode	T1	T2	T3	T4	T5	T6	T7	T8	T9	T10
ANSYS	0.6306	0.3406	0.2555	0.1789	0.1255	0.0957	0.0914	0.0913	0.0705	0.0609
MATLAB	0.6845	0.3414	0.2567	0.1790	0.1252	0.0952	0.0910	0.0905	0.0703	0.0605
Error	−8.56%	−0.24%	−0.47%	−0.08%	0.25%	0.57%	0.46%	0.91%	0.29%	0.64%

Table 8.5: One-dimensional wind velocity field of the cable-stayed bridge

Wind Field Number	Location	Direction	Simulation Point
2	Left tower	Longitudinal	5
4	Right tower	Longitudinal	5
6	Main girder	Vertical	23

According to the above three processes, the wind velocity time history can be obtained if the target power spectrum $[S^0(\omega)]$ is given. In this section, the Kaimal horizontal fluctuating wind velocity spectrum [248] is adopted as the target power spectrum. The main factors in the wind field simulation are as follows: the span $L = 225$ m, the effective height of the girder from the ground is $z = 20$ m, the surface roughness $z_0 = 0.03$ m, the average wind velocity at the girder is $U(z) = 30$ m/s, the number of simulation points $n = 23$, the space of simulation points is $\Delta = 10$ m, the upper limit frequency is $\omega_{up} = 2\pi$ rad/s, the number of frequency division is $N = 256$, and the sampling time interval is $\Delta t = 0.5$ s.

The wind field can be simplified into three independent one-dimension multivariable random wind velocity fields, as shown in Table 8.5. Twenty-three simulation points are distributed along the girder from the left to the right with a space of 10 m, 5 simulation points are distributed along the main tower from the bottom to the top with a space of 5 m, and the distribution of simulated points is shown in Fig. 8.8. According to the above theory, the wind velocity time history is obtained. Fig. 8.10 shows simulating results of point 1 and point 23.

8.2.3 Semiactive Control Strategy

In this example, the MR damper is installed in the cable-stayed bridge to control the longitudinal floating, the longitudinal displacement of the main tower, and the vertical bending displacement of the first mode of the bridge girder. According to the principle, the MR damper will be installed between the tower and the girder, as shown in Fig. 8.11, and the MR dampers are installed at points 48, 124, 192, and 286.

According to the previous intelligent control algorithm discussed in Chapter 2, Intelligent Control Strategies, the optimal control force of the MR damper can be calculated using the LQR optimal control algorithm. However, the MR damper cannot provide the calculated optimal control forces in any instant, thus the appropriate semiactive controlled strategy which fit the control of the bridge should be employed. Here, the idea of tristate control

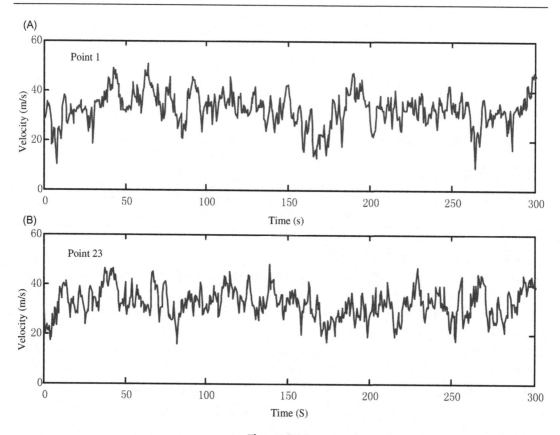

Figure 8.10
Simulation of the time-history curves of the wind velocity. (A) Time-history curve of wind velocity of point 1 and (B) time-history curve of wind velocity of point 23.

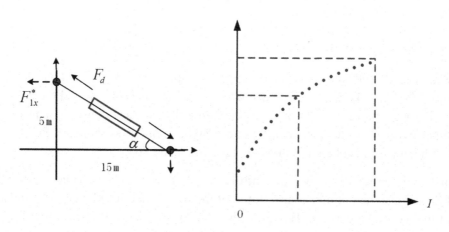

Figure 8.11
Installation of the MR damper and the adjusting of current.

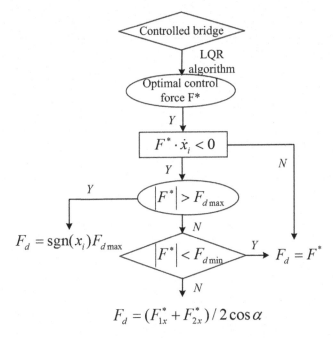

Figure 8.12
Diagram of the LQR tristate control strategy.

is adapted to adjust the inputting current, as seen in Eq. (8.6), which is similar to the semiactive controlled strategy discussed in Section 8.1.

$$F_d = \begin{cases} F^* & |F^*| < F_{dmin} \text{ or } F^* \cdot \dot{x}_i > 0 \\ (F^*_{1x} + F^*_{2x})/2\cos\alpha & F_{dmin} < |F^*| < F_{dmax} \text{ and } F^* \cdot \dot{x}_i < 0 \\ \text{sgn}(x_i) F_{dmax} & |F^*| > F_{dmax} \text{ and } F^* \cdot \dot{x}_i < 0 \end{cases} \quad (8.6)$$

where F_{dmin} and F_{dmax} are the control forces of MR dampers with no current and the magnetic saturation current. x_i is the displacement of the control point, F^* is the optimal control force, F^*_{1x} and F^*_{2x} are the longitudinal optimal control forces of the control points 1 and 2, respectively, and α is the installation angle, as shown in Fig. 8.11.

The corresponding schematic diagram of the control strategy is shown in Fig. 8.12. The control forces are changing with the dynamic responses of each time through adjusting the input current.

8.2.4 Results and Analysis

Based on the above modeling and control strategy, the dynamic responses of the controlled and uncontrolled bridges are calculated. Fig. 8.13(A) shows the longitudinal displacement response of the bridge. The maximum longitudinal displacement responses are 141.5 mm for the uncontrolled bridge, 46.9 mm for the optimal controlled bridge with a reduction of 66.85%,

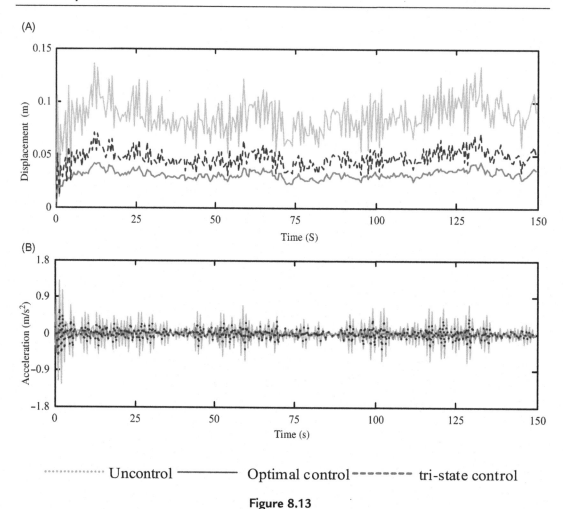

Figure 8.13
Dynamic responses of the main girder in the longitudinal direction. (A) Longitudinal displacement responses and (B) longitudinal acceleration responses.

and 71.0 mm for the tristate controlled bridge with a reduction of 49.82%. The tristate control strategy can effectively reduce the dynamic responses of longitudinal displacement responses of the bridge, although its control effect is inferior to the optimal control effect. Fig. 8.13(B) shows the acceleration responses of the bridge, it can be seen that the longitudinal acceleration responses are significantly reduced by using the MR damper. The acceleration peak value is 1.28 m/s^2 for the uncontrolled bridge, 0.62 m/s^2 for the controlled bridge with the tristate control strategy, and 0.42 m/s^2 for the controlled bridge with the optimal control strategy.

The longitudinal displacement peak value of the main tower of the uncontrolled, the LQR optimal controlled, and the tristate controlled bridges are calculated, and the results are listed in Table 8.6. It can be seen that the displacement peak value is significantly reduced by using MR dampers. The peak value of the uncontrolled bridge is 132.8 mm, the peak value of the LQR optimal controlled bridge is 47.8 mm with a reduction of 64.01%, and the

Table. 8.6: Peak vertical displacement of the of the tower under different control strategies

Peak Displacement (m) Coordinate (x, y)	Uncontrolled and LOR Tristate Control			Uncontrolled and LOR Control		
	Uncontrolled	Tristate Control	Control Effect	Uncontrolled	Optimal Control	Control Effect
(−62.5, 30)	0.1328	0.0722	45.63%	0.1328	0.0478	64.01%
(−62.5, 25)	0.1331	0.0734	44.85%	0.1331	0.0505	62.06%
(−62.5, 20)	0.1224	0.0682	44.28%	0.1224	0.0480	60.78%
(−62.5, 15)	0.0932	0.0523	43.88%	0.0932	0.0375	59.76%
(−62.5, 10)	0.0530	0.0299	43.58%	0.0530	0.0216	59.25%
(−62.5, 5)	0.0162	0.0092	43.21%	0.0162	0.0067	58.64%
(−62.5, 0)	0	0		0	0	

peak value of the tristate controlled bridge is 72.2 mm with a reduction of 45.63% compared with the uncontrolled bridge.

Although the angle of the MR damper is relatively small to provide enough force to control the longitudinal displacement, the MR damper can also provide a vertical force to control the vertical displacement of the middle span node of the girder. The vertical displacement and acceleration responses of the middle span node are calculated and the results are shown in Fig. 8.14.

It can be seen from Fig. 8.14 that the peak value of vertical displacement responses of the middle span node is 88.5 mm for the uncontrolled bridge, and 80.7 mm for the LQR optimal controlled bridge with a reduction of 8.8%, and 59.5 mm for the LQR tristate controlled bridge with a reduction of 32.77%. The peak value of vertical acceleration responses of the middle span node is 0.8 m/s^2 for the uncontrolled bridge, and 0.62 m/s^2 for the LQR optimal controlled bridge with a reduction of 22.5%, and 0.46 m/s^2 for the tristate controlled bridge with a reduction of 42.5%.

In order to show the control effect of the MR damper on the displacements of the main tower and girder, the peak values of sixteen points are extracted on the left main tower with an interval of 2 m, and the displacement envelop diagram of the left main tower is shown in Fig. 8.15. Similarly, the displacement envelop diagram of the girder is plotted by extracting the displacement of 46 points on the girder with an interval of 5 m, as shown in Fig. 8.16. It can be seen that the MR damper can significantly reduce the dynamic responses of the bridge structure.

8.3 Dynamic Analysis on Platform With MRE Devices

Many precision industrial and experimental processes cannot work accurately if the instruments are affected by external vibrations, since vibration sensitive components are used in modern mechanism, such as atomic force microscopes, space telescopes and interferometers. Generally, these precision equipment and optical instruments are usually placed on a platform, and the vibration energy will be transmitted to the instruments

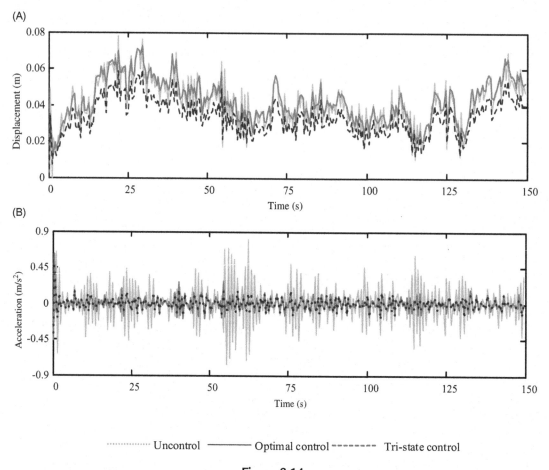

Figure 8.14
Vertical dynamic responses of the main girder (point 1). (A) Vertical displacement responses of the middle span and (B) vertical acceleration responses of the middle span.

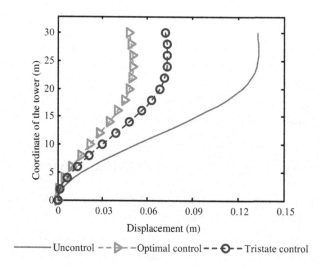

Figure 8.15
Envelop of the longitudinal displacement of the main tower.

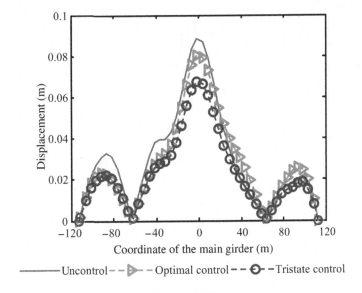

Figure 8.16
Envelop of the vertical displacement of the main girder.

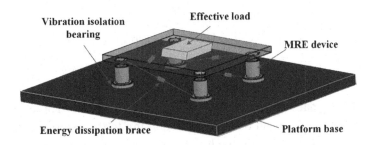

Figure 8.17
Vibration isolation scheme of the platform system.

through the platform. Therefore, vibration suppression measures should be taken into account on platforms. In this section, dynamic response analysis of a platform structure with MRE devices will be introduced.

8.3.1 Modeling and Parameters

Fig. 8.17 shows the vibration isolation scheme of the platform system, which includes vibration isolation bearing, MRE device, and energy dissipation brace. The vibration isolation bearing is made of energy-guzzling viscoelastic material, which is used to isolate the excitation with high frequency and dissipate the input energy, so it is installed on the top of the platform base and connects to the MRE device. The energy dissipation brace is

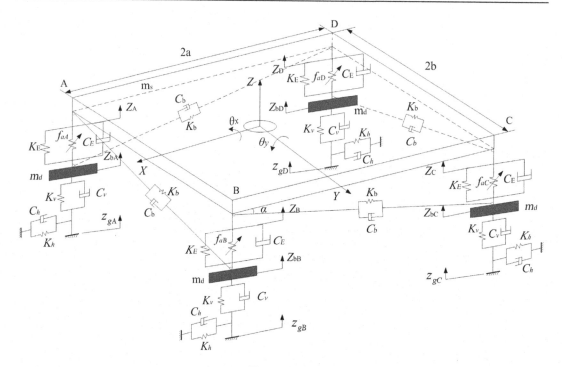

Figure 8.18
The simplified model of the platform structure.

also made of energy-guzzling viscoelastic material, and installed between the two legs to resist the in-plane torsional vibrations of the platform. The stiffness and energy dissipation properties of the MRE devices can be adjusted according to the external excitation and structural responses, so as to mitigate the low frequency vibration.

Based on the vibration mitigation mechanism of the platform system, the mathematical model of vibration isolation control system is established, as shown in Fig. 8.18. Considering the working environment and the motion state of the platform, the vibration isolation and mitigation system of the platform is simplified to a model with seven degree of freedom.

Parameters are specified as follows:

  z: the vertical displacement of platform centroid;
  θ_y: the angular displacement of pitching motion state of the platform;
  θ_x: the angular displacement of lateroversion motion state of the platform;
  z_A, z_B, z_C, z_D: the vertical displacements of platform corners A, B, C, D;
  z_{bA}, z_{bB}, z_{bC}, z_{bD}: the vertical displacements of the platform legs A, B, C, D;
  m_s: the effective mass of the platform;

m_d: the mass of platform legs;
K_E: the stiffness coefficient of MRE device;
$f_{aA}, f_{aB}, f_{aC}, f_{aD}$: the magnetic forces of MRE devices;
C_E: the damping coefficient of MRE device;
K_v, K_h: the vertical and horizontal stiffness coefficient of the vibration isolating bearing;
C_v, C_h: the vertical and horizontal damping coefficient of the vibration isolating bearing;
K_b, C_b: the stiffness and damping coefficient of the energy dissipation brace;
a, b: the size parameters of platform;
$Z_{gA}, Z_{gB}, Z_{gC}, Z_{gD}$: the input excitation of each leg.

Based on the mathematical model, the motion equation of the controlled platform structure can be expressed as

$$[M]\{\ddot{x}(t)\} + [C]\{\dot{x}(t)\} + [K]\{x(t)\} = \{P(t)\} - [B]\{f_d(t)\} \tag{8.7}$$

where $\{x(t)\} = \{x_1, x_2, x_3, x_4, x_5, x_6, x_7\}^T$, in which x_1, x_2, and x_3 separately are the vertical, pitching motion, and side tumbling motion displacement responses, and x_4, x_5, x_6, and x_7 separately are the displacement responses of platform legs. $[M] = \text{diag}(m_s, I_{yy}, I_{xx}, m_d, m_d, m_d, m_d)$ is the mass matrix of the control system. $[C]$ and $[K]$ separately are the damping and stiffness matrixes. $[B]$ is the position matrix of damper control force. $\{P(t)\}$ and $\{f_d(t)\}$ separately are the exciting force and damper control force. Introducing state vector $\{Z(t)\} = \{\{x(t)\}, \{\dot{x}(t)\}\}^T$, the above equation can be expressed in the form of state-space equations.

Combined with the features of the platform (mass and size) in this example, and on the basis of the preliminary analysis, the parameters of the platform structure control system are shown in Table 8.7.

Table 8.7: Parameters of the vibration control system of the platform

Parameter	Meaning of Parameter	Value
m_s	The effective mass of platform	300 kg
m_d	The mass of platform legs	30 kg
I_{xx}	The rotational inertia of the Y axis	1091.6 kg·m²
I_{yy}	The rotational inertia of the X axis	1150.3 kg·m²
K_v	The vertical stiffness coefficient of the vibration isolating bearing	1.5×10^6 N/m
C_v	The vertical damping coefficient of the vibration isolating bearing	0.9×10^5 N·s/m
K_E	The initial stiffness coefficient of MRE device	1.3×10^6 N/m
C_E	The initial damping coefficient of MRE device	3.8×10^4 N·s/m
K_b	The stiffness coefficient of the energy dissipation brace	0.5×10^6 N/m
C_b	The damping coefficient of the energy dissipation brace	3.0×10^4 N·s/m
$2a$	Long side size of the platform	1.2 m
$2b$	Short side size of the platform	1.0 m

In Eq. (8.7), $P(t)$ is a vector of the input excitation, which consists of the input excitations at the four supports of the platform. For structural dynamic analysis, the excitation data should be in the form of time-history input. In this example, the platform is thought to be rigid, so the time-history data of each input point (four supports of the platform) can be achieved from Eq. (8.8).

$$\begin{cases} z_A = z - a\theta_y - b\theta_x \\ z_B = z - a\theta_y + b\theta_x \\ z_C = z + a\theta_y + b\theta_x \\ z_D = z + a\theta_y - b\theta_x \end{cases} \tag{8.8}$$

8.3.2 Semiactive Control Strategy

For the platform structure with control devices, as shown in Fig. 8.17, when the parameters and excitation inputs of the motion equation are determined, dynamic analysis can be conducted by using MATLAB.

The optimal control force of the MRE device can be calculated using the LQR optimal control algorithm. However, the MRE device cannot provide the calculated optimal control forces in any instantaneous, thus an appropriate semiactive control strategy of the MRE device (Section 4.7) is designed, seen in Eq. (8.9).

$$\begin{cases} f_{MRE} = 0 & I = 0 & \text{when } |S_{d,p}| \leq a \\ f_{MRE} = f_{MRE,med} & I = \alpha I_{max} & \text{when } |S_{d,p}| > a \text{ and } 0 < \left|\dfrac{S_{d,MRE}}{h_v}\right| \leq 0.05 \\ f_{MRE} = f_{MRE,max} & I = I_{max} & \text{when } |S_{d,p}| > a \text{ and } 0.05 < \left|\dfrac{S_{d,MRE}}{h_v}\right| \leq 0.1, \text{ or } \left|\dfrac{S_{d,MRE}}{h_v}\right| > 0.1 \end{cases} \tag{8.9}$$

where $|S_{d,p}|$ is the vertical displacement absolute value of the platform centroid, a is the control target value of the platform centroid vertical displacement, $|S_{d,MRE}|$ is the stroke of MRE device, h_v is the thickness of MRE, and α is an adjustment coefficient of the control current. When the vertical displacement response meets the control requirement, i.e., $|S_{d,p}| \leq a$, the current value is 0. When the vertical displacement response cannot meet the control requirement and the maximum displacement response

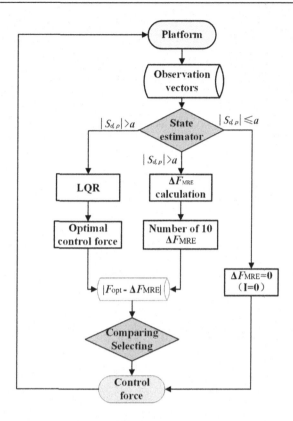

Figure 8.19
The control strategy process of platform structure.

is less than the stroke of MRE device, the output force of MRE device can be controlled to close to the optimal control force by changing the current value. When the vertical displacement response is in the last state, the current value achieves as the maximum value.

Based on the above control strategy, the current of MRE device needs to be chosen to realize the control effect, therefore the core idea for choosing current is dividing the work current into many states, and each state of working current corresponds with the output force value of MRE device. Fig. 8.19 shows the control strategy of the controlled platform structure, in which the response information about the controlled platform can be observed by sensors at any time, then the current is chosen according to the control law, which needs to be close to the optimal control force calculated by LQR method.

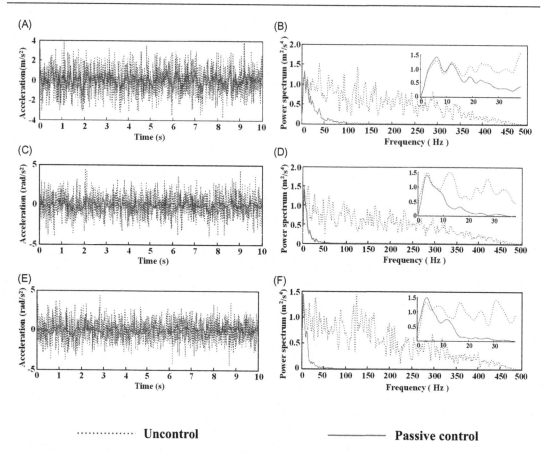

Figure 8.20

The acceleration response analysis of the platform with passive control. (A) The vertical acceleration time-history curves, (B) the vertical acceleration power spectrum curves, (C) the pitching angular acceleration time-history curves, (D) the pitching angular acceleration power spectrum curves, (E) the lateroversion angular acceleration time-history curves, and (F) the lateroversion angular acceleration power spectrum curves.

8.3.3 Results and Analysis

Based on the above modeling and control strategy, the dynamic responses of the controlled and uncontrolled platform structures are calculated. Fig. 8.20 shows the acceleration responses of the control system under passive control (the MRE device without current input, i.e., $I = 0$). It can be seen that the acceleration responses of three directions under passive control obviously decreased. The power spectrum curves show that the dynamic responses of the passive controlled platform are reduced effectively compared to the uncontrolled platform when the excitation frequency is larger than

Table 8.8: Analysis results in frequency domain (passive control)

Frequency		0–10 Hz			10–30 Hz			30–50 Hz		
Control Index		Z	θ_y	θ_x	Z	θ_y	θ_x	Z	θ_y	θ_x
Uncontrol	Peak value	1.21	1.49	1.28	1.17	1.50	1.27	1.51	1.14	1.37
Passive control	Peak value	1.32	1.4	1.43	1.11	0.75	0.72	0.38	0.11	0.09
	Reduction rate	−9.3%	6.1%	−11.8%	5.1%	50%	43.3%	74.9%	90.5%	93.4%

50 Hz, however, when the excitation frequency is in the scope of 0–10 Hz, the control effect is not obvious.

Table 8.8 shows the detail analysis results in frequency domain. It can be seen that when the frequency is in the scope of 30–50 Hz, the peak value of the power spectrum in Z direction, θ_y direction, and θ_x direction of the platform separately decreased from 1.51 m^2/s^4, 1.14 rad^2/s^4, and 1.37 rad^2/s^4 to 0.38 m^2/s^4, 0.11 rad^2/s^4 and 0.09 rad^2/s^4, respectively, and the corresponding reduction rates separately are 74.9%, 90.5%, and 93.4%. For the frequency 0–10 Hz, the reduction rates in three directions are −9.3%, 6.1%, and −11.8% separately, which means that the responses of the platform with passive control in the low frequency cannot be well controlled.

Fig. 8.21 shows the analysis results of the semiactive controlled platform with the control strategy mentioned above, it can be seen that the acceleration responses in three directions are all well controlled, especially the peak value of acceleration power spectrum in low frequency (0–10 Hz). Compared to the passive control, the semiactive control with the appropriate control strategy can reach the better control effect in the whole spectrum (0–500 Hz).

Table 8.9 shows the detail analysis results in frequency domain. It can be seen that when the frequency is in the scope of 0–10 Hz, the peak value of the power spectrum in Z direction, θ_y direction, and θ_x direction of the platform separately decrease from 1.21 m^2/s^4, 1.49 rad^2/s^4, and 1.28 rad^2/s^4 to 0.75 m^2/s^4, 0.70 rad^2/s^4, and 0.78 rad^2/s^4, and the corresponding reduction rates separately are 38%, 53%, and 39.1%.

8.4 SIMULINK Analysis Example

In Section 7.3, a simple example is used to introduce how the SIMULINK is operated. In this section, there are two SIMULINK examples to be introduced: one is a SIMULINK model of the structure without dampers, and the other is a SIMULINK model of the structure with MR dampers controlled by using the fuzzy controller and the neural network prediction model.

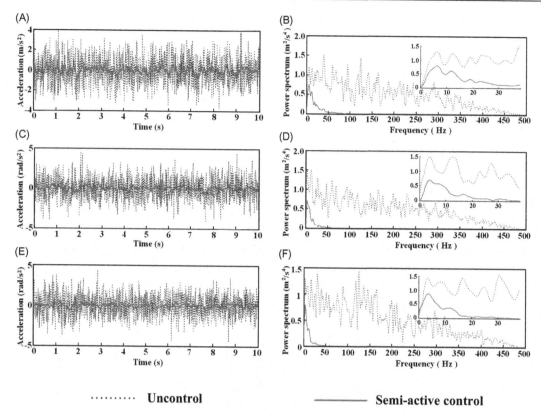

·········· **Uncontrol** ——— **Semi-active control**

Figure 8.21
The acceleration response analysis of the platform with semiactive control. (A) The vertical acceleration time-history curves, (B) the vertical acceleration power spectrum curves, (C) the pitching angular acceleration time-history curves, (D) the pitching angular acceleration power spectrum curves, (E) the lateroversion angular acceleration time-history curves, and (F) the lateroversion angular power spectrum curves.

Table 8.9: Analysis results in frequency domain (semiactive control)

Frequency		0–10 Hz			10–30 Hz			30–50 Hz		
Control Index		z	θ_y	θ_x	z	θ_y	θ_x	z	θ_y	θ_x
Uncontrol	Peak value	1.21	1.49	1.28	1.17	1.50	1.27	1.51	1.14	1.37
Semiactive control	Peak value	0.75	0.70	0.78	0.57	0.45	0.34	0.13	0.08	0.05
	Reduction rate	38.0%	53.0%	39.1%	51.3%	70.0%	73.2%	91.4%	93.0%	96.4%

8.4.1 The SIMULINK Example of the Structure Without Dampers

In this example, a five-story steel frame structure without dampers is modeled and simulated by SIMULINK. The model story height vector is $\{h\} = \{3.9, 3.3, 3.3, 3.3, 3.3\}^T$ m.

The structural lumped mass matrix [M], the structural stiffness matrix [K], and the structural damping matrix [C] can be gained.

$$[M] = \begin{bmatrix} 2.60 & 0 & 0 & 0 & 0 \\ 0 & 2.30 & 0 & 0 & 0 \\ 0 & 0 & 2.30 & 0 & 0 \\ 0 & 0 & 0 & 2.30 & 0 \\ 0 & 0 & 0 & 0 & 1.96 \end{bmatrix} \times 10^4 \text{ kg},$$

$$[K] = \begin{bmatrix} 4.38 & -2.32 & 0 & 0 & 0 \\ -2.32 & 4.64 & -2.32 & 0 & 0 \\ 0 & -2.32 & 4.64 & -2.32 & 0 \\ 0 & 0 & -2.32 & 4.64 & -2.32 \\ 0 & 0 & 0 & -2.32 & 2.32 \end{bmatrix} \times 10^7 \text{ N/m},$$

$$[C] = \begin{bmatrix} 2.15 & -1.06 & 0 & 0 & 0 \\ -1.06 & 2.25 & -1.06 & 0 & 0 \\ 0 & -1.06 & 2.25 & -1.06 & 0 \\ 0 & 0 & -1.06 & 2.25 & -1.06 \\ 0 & 0 & 0 & -1.06 & 1.17 \end{bmatrix} \times 10^5 \text{ N·s/m},$$

El Centro earthquake with 200 gal acceleration amplitude is selected, and the sampling time is 0.02 s. According to Eqs. (2.4) and (2.5), the state-space equation of the structure is given as:

$$\begin{cases} \{\dot{Z}(t)\} = [A]\{Z(t)\} - [D_0]\{\Gamma\}\ddot{x}_g(t) \\ \{Y\} = [E_0]\{Z(t)\} \end{cases} \tag{8.10}$$

According to the Eq. (8.10), the SIMULINK model of the structure is built, as shown in Fig. 8.22. The "Xg" block is the input module of El Centro earthquake with 200 gal

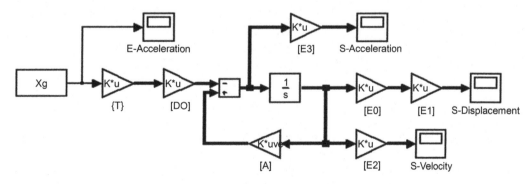

Figure 8.22
The SIMULINK model of the structure.

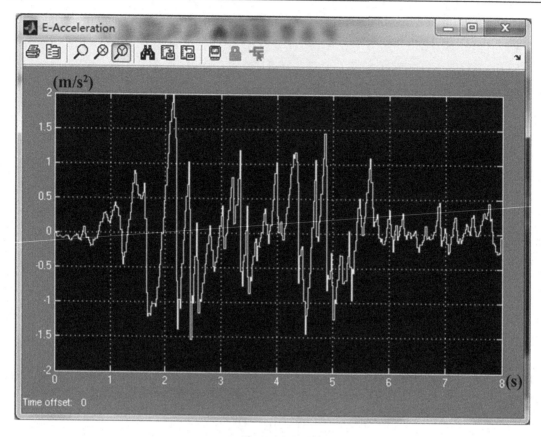

Figure 8.23
El Centro earthquake wave.

acceleration amplitude. The "[A]" block, "[D0]" block, "[E0]" block, and "{T}" block produce the system matrix $[A]$, the input matrix $[D_0]$, the output matrix $[E_0]$, and the vector $\{\Gamma\}$, respectively. The "[E1]" block, "[E2]," and "[E3]" are the matrices to be used to extract the displacement, velocity, and acceleration responses of the first, third, and the top floors of the structure, respectively. There are four scopes, "E-Acceleration," "S-Displacement," "S-Velocity," and "S-Acceleration," to display the El Centro earthquake wave, the displacement, velocity, and acceleration responses of the first, third, and the top floors of the structure, respectively, as shown in Figs. 8.23–8.26. (Note: all units adopt the international system of units.)

8.4.2 The SIMULINK Example of the Controlled Structure

In this example, a five-story steel frame structure with two MR dampers placed in parallel in the first floor is simulated using SIMULINK, in which the structural parameters are the

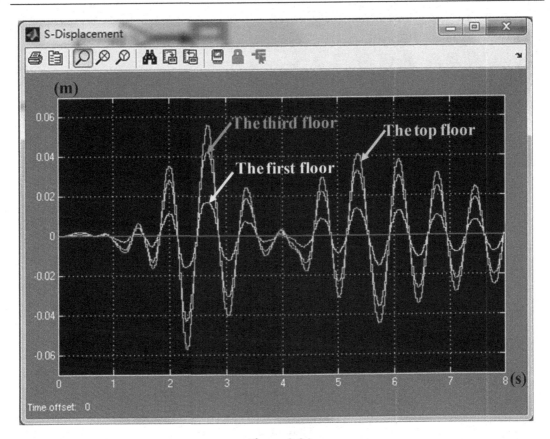

Figure 8.24
The displacement responses of the structure.

same as the model discussed in Section 8.4.1. The common MR dampers are adopted. The viscosity η is 0.9 Pa·s. The effective length of the piston L is 400 mm, the gap D_h is 2 mm, and the inner diameter of the cylinder D is 100 mm. A fuzzy control strategy is proposed to produce the control currents of the MR damper, and a neuro network forecasting model of the steel frame structure is developed to predict the seismic responses of the structure with MR dampers, including the displacement and velocity responses of the first floor. El Centro earthquake with 200 gal acceleration amplitude is selected as the input excitation, and the sampling time is 0.02 s.

For frame structures, MR dampers are usually placed between the chevron braces as shown in Fig. 8.27, and the state-space equation of the controlled structure is given as Eq. (2.5). For the MR dampers, the most frequently referred Bingham model [197] is used to simulate

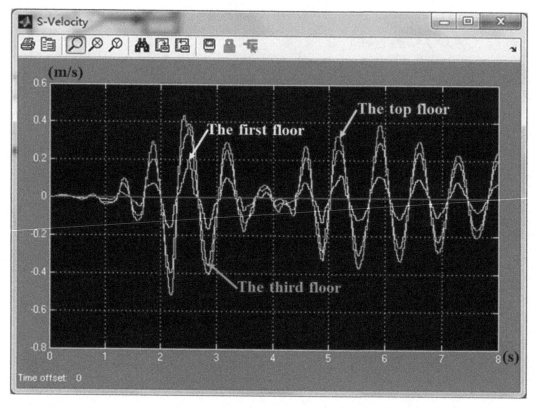

Figure 8.25
The velocity responses of the structure.

the properties of the MR dampers, which includes a friction element in parallel to a viscous element as shown in Fig. 8.28, and a relation between the stress and strain rate is expressed as:

$$\tau = \tau_y \,\text{sgn}(\dot{\gamma}) + \eta\dot{\gamma} \quad (8.11)$$

where τ is the shear stress in the fluid, η is the Newtonian viscosity which is independent of the applied magnetic field, $\dot{\gamma}$ is the shear strain rate and τ_y is the yield shear stress controlled by the applied magnetic field. Based on Eq. (8.11), Phillips [233] derived the force–displacement relationship for MR dampers:

$$f_d = f_c \,\text{sgn}[\dot{u}(t)] + c_0 \dot{u}(t) \quad (8.12)$$

where $f_c = \frac{3L_d A_p \tau_y}{h_d}$ is the frictional force, $c_0 = \frac{12\eta L_d A_p^2}{\pi D h_d^3}$ is the damping coefficient, L_d is the length of the piston, A_p is the cross-sectional area of the piston, D is the inner diameter of the cylinder, h_d is the gap between the piston and the cylinder, $u(t)$ is the relative displacement of the piston to the cylinder, and τ_y is the function of the applied magnetic

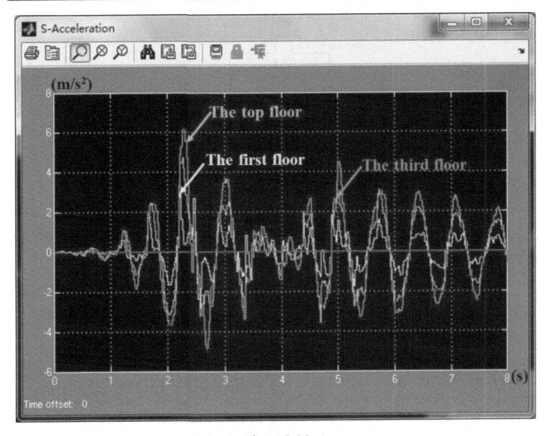

Figure 8.26
The acceleration responses of the structure.

Figure 8.27
Schematic of smart structure.

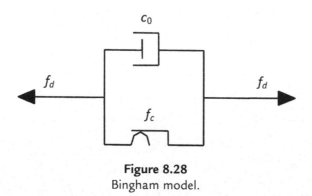

Figure 8.28
Bingham model.

field, which means τ_y is the function of control currents I_c. Xu [249] proposed the relation between τ_y and the control I_c of the MR damper as follows:

$$\tau_y = A_1 e^{-I_c} + A_2 \ln(I_c + e) + A_3 I_c \qquad (8.13)$$

where $A_1 = -11374$, $A_2 = 14580$, and $A_3 = 1281$ are coefficients relative to the property of the MR fluid in the MR damper, and e is a constant.

A fuzzy controller is designed to control the working state of MR dampers. The two input variables are the earthquake acceleration excitation and the structural first floor displacement predicted by the neural network. The output variable is control current of MR dampers. The selection of the input variables of the fuzzy controller is based on the following two reasons: (1) The earthquake acceleration excitation influences on seismic responses of structures directly; (2) inter-story drifts are limited in seismic code, e.g., $\Delta u \leq h/550$ (h is the story height) is specified in the seismic code in China, so displacement responses are important parameters reflecting the control effects.

The basic domain of the earthquake acceleration is determined in accordance with the amplitude of the input acceleration. For an unknown earthquake wave beforehand, the basic domain can usually be determined as 0–10 m/s^2, otherwise, the basic domain can be determined by the referenced range according to the limit of the magnitude of earthquake acceleration [250]. In this example, the 0.2 g El Centro earthquake wave is adopted as earthquake input, and then the basic domain is determined as 0–4 m/s^2. The elastic limit of story drift is $h/550$ for frame structures according to the seismic code in China, where h is the story height. Therefore for the displacement response, the basic domain is 0–$h/550$. The basic domain of control currents of MR dampers is 0–2 A, same as the working current of MR dampers. According to basic domains of the earthquake acceleration excitation, the first floor displacement and control currents of MR dampers, fuzzy domains are determined as 0–4, 0–6 and 0–2, respectively. Accordingly, the corresponding quantification factors or proportion factors are determined as: the quantification factor of earthquake wave is $K_a = 2$, the quantification factor of the first floor displacement of structure is $K_d = 104$, the proportion factor of control currents of MR dampers is $K_c = 1$.

The earthquake acceleration, the structural first floor displacement and control currents of MR dampers may be divided into five grades, i.e., {VS (very small), S (small), M (middle), B (big), VB (very big)}, and their membership function curves are plotted in Fig. 8.29. Fuzzy rules of the fuzzy controller are listed in Table 8.10.

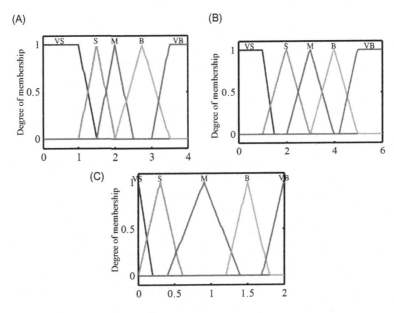

Figure 8.29
The membership function curves of the inputs and outputs of the fuzzy controller. (A) Fuzzy domain of the earthquake, (B) fuzzy domain of the displacement, and (C) fuzzy domain of the current.

Table 8.10: Fuzzy rules of the fuzzy controller

Displacement	Acceleration				
	VS	S	M	B	VB
VS	VS	VS	VS	VS	S
S	S	S	S	M	M
M	S	M	B	B	VB
B	S	B	B	VB	VB
VB	B	VB	VB	VB	VB

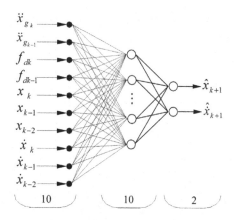

Figure 8.30
The neural network architecture.

A three-layer feed-forward neural network model is trained to predict structural dynamic responses to resolve the time-delay problem of the structure with MR dampers. The neural network model consists of an input layer, a hidden layer, and an output layer, as shown in Fig. 8.30. The input variables of the neural network are the $(k-1)$th and kth time earthquake accelerations ($\ddot{x}_{g_{k-1}}, \ddot{x}_{g_k}$), the $(k-1)$th and kth time control forces (f_{dk-1}, f_{dk}) and the $(k-2)$th, $(k-1)$th, and kth time seismic response including the displacement and velocity of the first floor ($x_{k-2}, x_{k-1}, x_k, \dot{x}_{k-2}, \dot{x}_{k-1}, \dot{x}_k$). The output variables of the neural network are the next time-step displacement response of the first floor $\hat{x}_{k+1}$, and the next time-step velocity response of the first floor $\hat{\dot{x}}_{k+1}$. The predicted value $\hat{x}_{k+1}$ is one input variable of the fuzzy controller, and the predicted value $\hat{\dot{x}}_{k+1}$ is one of variables applied to gain control forces of MR dampers installed the first floor of the structure.

Based on the analysis and design, the SIMULINK model of the intelligent control system of the controlled steel frame structure is built in accordance with the SIMULINK operations in Section 7.3, as shown in Fig. 8.31. The "NNET" block is the neural network forecasting model used to predict seismic responses of the structure with MR dampers, including the displacement and velocity of the first floor. The "fuzzy logic controller" block is the fuzzy controller used to produce controlled current of MR dampers. The "Ka" block produces the quantification factor of earthquake wave value, $K_a = 1$; the "Kd" block produces the quantification factor of the displacement response of the first floor structure, $K_d = 104$; the "Kc" block produces the proportion factor of control currents of MR dampers value, $K_c = 1$. The "MR damper" block is the subsystem simulating the nonlinear behavior of MR dampers, as shown in Fig. 8.32. The "E-Acceleration" block is an earthquake excitation source. The "E1" block is a matrix used to extract the first floor displacement from dynamic responses of the structure; the "C1" block is a matrix used to extract the first floor velocity from dynamic responses of the structure; the "C2" block is a matrix used to extract the predicted displacement of the first floor of the structure from the values predicted by the neural network forecasting

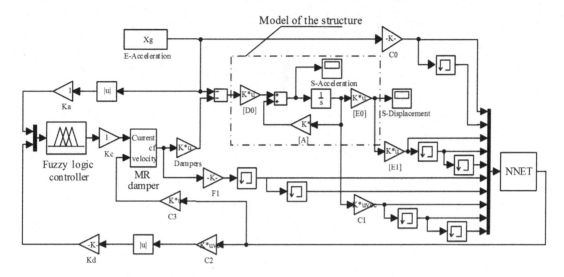

Figure 8.31
The SIMULINK model of the controlled structure.

model; the "C3" block is a matrix used to extract the predicted velocity of the first floor of the structure from the values predicted by the neural network forecasting model. The "S-Displacement" block displays displacement responses of the fuzzy control system.

Fig. 8.33(A) and (B) shows the displacement and acceleration responses of the top floor of the fuzzy controlled and uncontrolled structures. It can be seen from Fig. 8.33 that both the displacement and acceleration responses of the fuzzy controlled structure are reduced effectively, especially for the displacement responses. The maximum displacement response of the top floor in the fuzzy controlled structure is 45.60 mm, by the reduction of 20.74% compared with that of the uncontrolled structure, 57.53 mm. At the same time, the maximum acceleration response of the top floor in the fuzzy controlled structure is 5.27 m/s^2, by the reduction of 14.31% compared with that of the uncontrolled structure, 6.15 m/s^2. Notice that, the seismic responses of the fuzzy controlled structure slightly increase at the early stage of the simulation. One contributing factor for this phenomenon is the deviation existing in the neural network forecasting model in the early simulation. The neural network is trained on line by collecting every time-step seismic responses of the structure with MR dampers, however, at the early stage, it is lack of training data. Hence, the predicted displacement and velocity responses are not accurate enough, and the error of the control force of MR dampers exists inevitably, which may result in the magnification of dynamic responses. Nevertheless, the magnification does not affect the control effect of the fuzzy control strategy due to that the period of magnified dynamic responses is very short.

For the sake of illuminating the efficiency of the fuzzy control strategy, the structural seismic responses under the fuzzy control strategy are compared with those under the

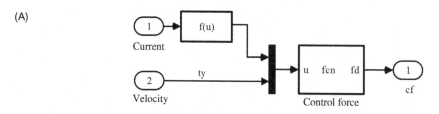

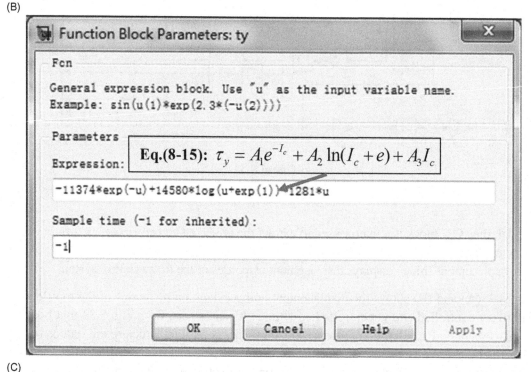

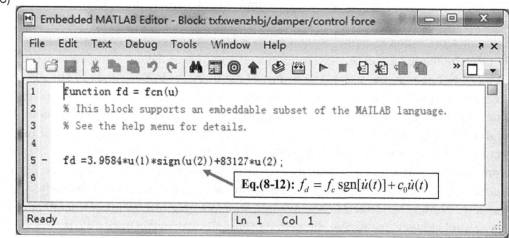

Figure 8.32
The SIMULINK model of the MR damper. (A) The subsystem of the MR damper, (B) setting of the "ty" block, and (C) setting of the "control force" block.

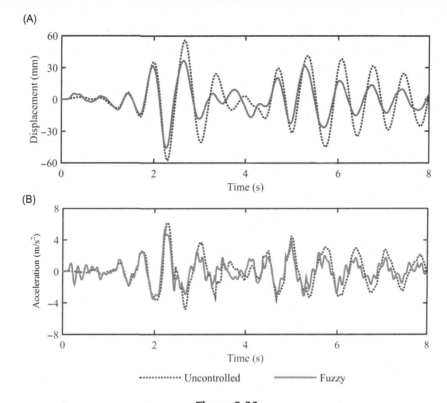

Figure 8.33

The top floor responses comparison between the fuzzy controlled structure and the uncontrolled structure. (A) The displacement response and (B) the acceleration response.

passive-off and passive-on control strategy. The passive-off and passive-on control strategies belong to the passive mode, i.e., MR dampers are held at 0 A in the passive-off control strategy and the maximum current level (2 A) in the passive-on controlled strategy. Fig. 8.34(A) and (B) shows the displacement and acceleration responses of the top floor among the fuzzy controlled structure and the passive-off controlled structure. Fig. 8.35(A) and (B) shows the displacement and acceleration responses of the top floor of the fuzzy controlled structure and the passive-on controlled structure. It can be seen from Fig. 8.34 that the maximum displacement response of the top floor in the fuzzy controlled structure is 45.60 mm, by the reduction of 14.69% compared with that of the passive-off controlled structure, 53.45 mm. The maximum acceleration response of the top floor in the fuzzy controlled structure is 5.27 m/s^2, by the reduction of 7.71% compared with that of the passive-off controlled structure, 5.71 m/s^2. It is easy to comprehend that the control effect of the fuzzy control strategy is better than that of the passive-off control strategy, because control forces of MR dampers under the fuzzy control strategy are stronger and more appropriate than those under the passive-off control strategy. It can be seen from Fig. 8.35 that the maximum displacement response of the top floor in the fuzzy controlled structure is

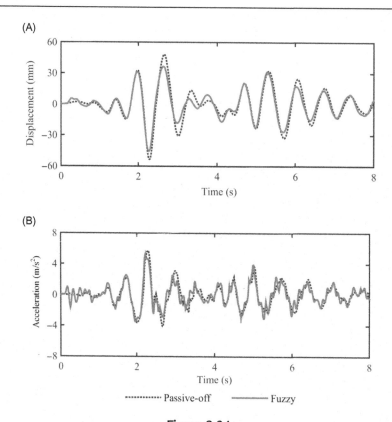

Figure 8.34
The top floor responses comparison between. (A) The displacement response and (B) the acceleration response.

45.60 mm, by the reduction of 8.19% compared with that of the passive-off controlled structure, 53.45 mm. The maximum acceleration response of the top floor in the fuzzy controlled structure is 5.27 m/s^2, by the reduction of 22.39% compared with that of the passive-on controlled structure, 6.79 m/s^2. Notice that the displacement responses under the fuzzy control strategy are slightly smaller than those under the passive-on control strategy; while the acceleration responses under the fuzzy control strategy are obviously smaller than those under the passive-on control strategy. The largest control forces produced by MR dampers usually donot benefit to control the acceleration under the passive-on control strategy. Nevertheless, the fuzzy control strategy can produce appropriate control forces to reduce the structural seismic responses. Apparently, choosing the passive-on strategy that produces the largest damping forces may not always be the most effective approach to protect the structure.

Compared with the commonly used bistate control strategy, how about the fuzzy control strategy? The bistate control strategy means that control currents of MR dampers can be turned on or off in real time. If the structure is moving away from its equilibrium position, control

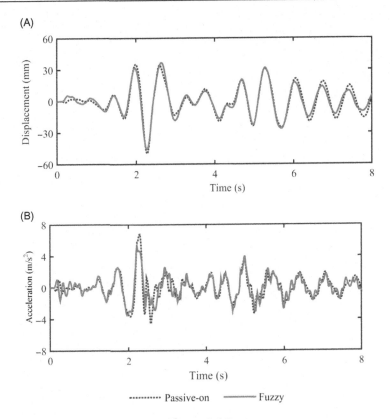

Figure 8.35
The top floor responses comparison between. (A) The displacement response and (B) the acceleration response.

currents of MR dampers are turned on and reach the maximum value. If the structure is returning to its equilibrium position, control currents of MR dampers are turned off and reach the minimum value. In this study, the maximum working current of MR dampers is 2 A, and the minimum current is 0 A under the bistate control method. Fig. 8.36(A) and (B) shows the displacement and acceleration responses of the top floor of the fuzzy controlled structure and the bistate controlled structure. It can be shown that the maximum displacement response of the top floor in the fuzzy controlled structure is 45.60 mm, by the reduction of 16.76% compared with that of the bi-state controlled structure, 54.78 mm. The maximum acceleration response of the top floor in the fuzzy controlled structure is 5.27 m/s^2, by the reduction of 30.84% compared with that of the bistate controlled structure, 7.62 m/s^2. The numerical comparison results illustrate that the bistate control strategy is not an ideal control strategy although it is easy to realize. The main reason is that the bistate control strategy is a rough control strategy which cannot generate proper control currents of MR dampers. The fuzzy control strategy is a fine control strategy having the strong robustness and can determine appropriate control currents of MR dampers.

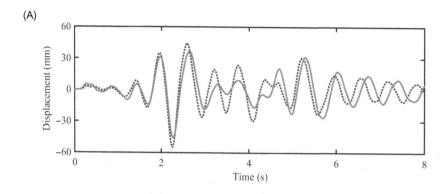

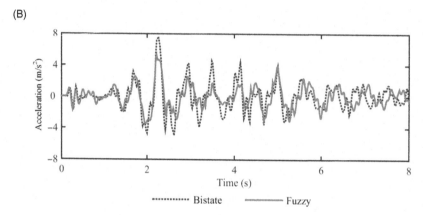

Figure 8.36
The top floor responses comparison between. (A) The displacement response and (B) the acceleration response.

The maximum displacement and acceleration responses comparison of each floor under different control strategies are given in Fig. 8.37(A) and (B). It can be seen that the fuzzy control strategy is the most effective control strategy among the above control strategies. Since the passive-on control strategy adopts the maximum control currents of MR damper, its earthquake mitigation effect for the structural displacement response is good, especially for the maximum displacement responses of the second floor (20.74 mm) and the third floor (34.27 mm), which are smaller than those of the fuzzy control strategy (24.40 mm, 35.15 mm). However, the maximum acceleration responses under the passive-on control strategy are larger than those under the fuzzy control strategy due to the large control forces of MR dampers, which means providing the large equivalent stiffness to the structure. The maximum acceleration responses are increased due to the inappropriate currents of MR dampers under the bistate control strategy, moreover, the earthquake mitigation effect of the bistate control strategy for the structural displacement responses is also worse than that of the fuzzy control strategy. Obviously, the control effect of the fuzzy control strategy is better than the passive-on control strategy and the bistate control strategy as a whole,

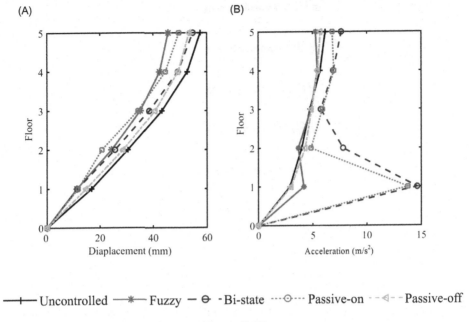

Figure 8.37
The maximum responses comparison of each floor under different control strategies.
(A) The maximum displacement response and (B) The maximum acceleration response.

especially for the acceleration responses. For example, the maximum acceleration response of the second floor in the fuzzy controlled structure is 3.68 m/s², by the reduction of 24.12% compared with that of the passive-on controlled structure, 4.85 m/s², and by the reduction of 52.76% compared with that of the bistate controlled structure, 7.79 m/s².

8.5 Particle Swarm Optimization Control Example

In this example, a five-story structure with five MR dampers (one MR damper is installed on each floor, as shown in Fig. 8.38) analyzed under the Particle Swarm Optimization (PSO) algorithm, the passive-on control algorithm, and the passive-off control algorithm, respectively.

8.5.1 Structural and Damper Parameters

The model structure is same as the model in Section 8.4.1. The common MR dampers are adopted. The MR fluids viscosity is 1.5 Pa·s. The effective length of the piston L is 844.3 mm, the inner diameter of the cylinder D is 203 mm, the piston cross-sectional area A_p is 1.09×10^4 mm², and the gap h is 2.06 mm. El Centro earthquake with 200 gal acceleration amplitude is selected, and the sampling time is 0.02 s.

In the PSO optimization control of the frame structure with MR dampers, multiobjective optimization control method is adopted. The objective function is the combination of the

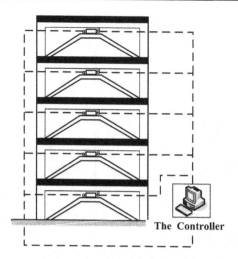

Figure 8.38
The structure with MR dampers.

displacement responses and the acceleration responses of the controlled structure, so the fitness function is obtained:

$$\begin{cases} \min F(t) \\ F(t) = \alpha f_1(t) + \beta f_2(t) \end{cases} \quad (8.14)$$

$$f_1(t) = \frac{|z_n|}{\max z_n} \quad (8.15)$$

$$f_2(t) = \left(\frac{|\ddot{z}_1|}{\max \ddot{z}_1} + \frac{|\ddot{z}_2|}{\max \ddot{z}_2} + \cdots + \frac{|\ddot{z}_n|}{\max \ddot{z}_n}\right)/n \quad (8.16)$$

where $f_1(t)$ and $f_2(t)$ are the objective function of the displacement responses and the acceleration responses of the structure with MR dampers, respectively. α and β are the weight coefficients, and $\alpha + \beta = 1$. In this example, $\alpha = 0.7$ and $\beta = 0.3$. t is the time variable. z_n and $\ddot{z}_n$ are the displacement response and the acceleration response of the nth floor of the structure, respectively. max z_n and max $\ddot{z}_n$ are the allowable maximum displacement response and acceleration response of the nth floor of the structure, respectively. According to the requirement of the building antiseismic design and building codes, when the structure is under the elastic state, the allowable maximum displacement response of the structure is $h/550$ (h is the story height of the structure); when the structure is under the elastoplastic state, the allowable maximum displacement response of the structure is $h/50$.

8.5.2 The PSO Optimization Control

In this example, the PSO algorithm with constriction factor is used to update the velocity and position of the particle, and the equations are shown as Eqs. (2.146) and (2.147).

The parameters are $c_1 = 2.8$, $c_2 = 1.3$, $\varphi = c_1 + c_2$. The particle swarm size is 30, and the number of the dimension in the problem space is 5.

In the PSO optimization control of the frame structure with MR dampers, the termination conditions of the PSO algorithm are the displacement and acceleration responses of the structure are less than the allowable value in the antiseismic design codes. That is, when the displacement and acceleration responses of the structure are less than the preset max z_n and max $\ddot{z}_n$, the optimization program will be terminated and the control current will be outputted.

8.5.3 Results and Analysis

For the sake of illuminating the efficiency of the PSO algorithm, the structure seismic responses under the PSO algorithm are compared with those of the uncontrolled, the passive-off and passive-on control structures. Fig. 8.39(A) and (B) shows the displacement and acceleration responses of the top floor of the PSO controlled structure and the uncontrolled structure. Fig. 8.40 (A) and (B) shows the displacement and acceleration responses of the top floor of the PSO controlled structure and the passive-off controlled structure. Fig. 8.41 (A) and (B) shows the displacement and acceleration responses of the top floor of the PSO controlled structure and the passive-on controlled structure.

It can be seen from Fig. 8.39 that the displacement responses of the PSO controlled structure are reduced effectively. The maximum displacement response of the top floor in the PSO controlled structure is 19.3 mm, by the reduction of 66.43%, compared with that of the uncontrolled structure, 57.5 mm. For the acceleration responses, the PSO algorithm can improve the earthquake mitigation effect slightly. The maximum acceleration response of the top floor in the PSO controlled structure is 5.90 m/s^2, by the reduction of 10.58% compared with that of the uncontrolled structure, 6.60 m/s^2.

As shown in Fig. 8.40, compared with the passive-off controlled structure, the displacement responses of the PSO controlled structure are reduced effectively. The maximum displacement response of the top floor in the PSO controlled structure is 19.3 mm, by the reduction of 63.03%, compared with that of the passive-off controlled structure, 52.2 mm. For the acceleration responses, the PSO control stratetgy is not better than the passive-off control strategy. The maximum acceleration response of the top floor in the PSO controlled structure is 5.90 m/s^2, by the magnification of 2.63% compared with that of the passive-off controlled structure, 5.75 m/s^2.

It can be seen from Fig. 8.41 that the PSO control algorithm is inferior to the passive-on control algorithm when the control objective is the displacement response.
For the acceleration responses, the PSO controlled structure is much better than the passive-on controlled structure. The maximum acceleration response of the top floor in the PSO controlled structure is 5.90 m/s^2, by the magnification of 17.11% compared with that of the passive-on controlled structure, 5.75 m/s^2.

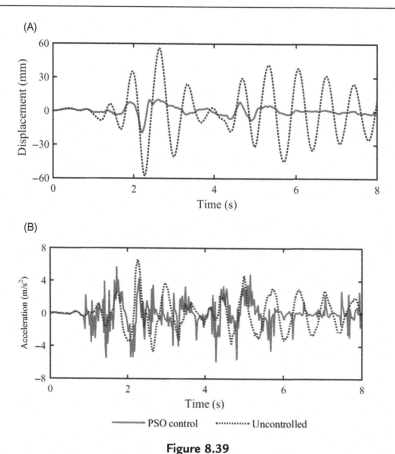

Figure 8.39
The top floor responses comparison between the PSO controlled structure and the uncontrolled structure. (A) The displacement response and (B) the acceleration response.

The maximum displacement and acceleration responses comparison of each floor under different control strategies are given in Fig. 8.42(A) and (B). It can be seen that the control effect of the maximum displacement response of each floor in the PSO controlled structure is inferior to that of passive-on controlled structure. Nevertheless, the control effect of structural displacement responses is significant, and the minimum reduction of the maximum displacement of each floor can reach 53.57% of the corresponding value of the uncontrolled structure, which meets the requirement of the building antiseismic design and building codes. For the maximum acceleration response of each floor, structural acceleration response increases under the passive-on control. This phenomenon can be explained that the working current of the MR damper is always the maximum current (2 A) for the passive-on controlled structure, which results in the increases of the structural stiffness unreasonably. The PSO control algorithm can reduce the maximum acceleration response of each floor effectively, this is because the PSO control algorithm adopts optimal control current as possible based on the

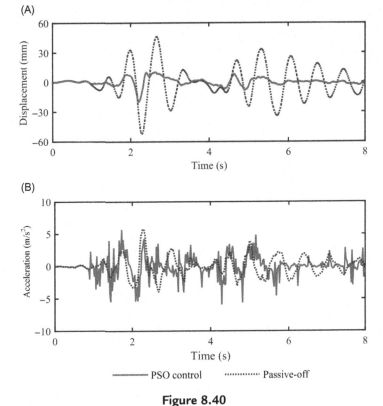

Figure 8.40
The top floor responses comparison between the PSO controlled structure and the passive-off controlled structure. (A) The displacement response and (B) the acceleration response.

fitness function and requirements real time. So, the displacement and acceleration responses of the structure can be controlled effectively.

8.6 Active Control Example

In this section, there is an active control example to be introduced. As mentioned in Chapter 3, Active Intelligent Control, active tendon system (ATS) is an effective and common way of active control. So, ATS is taken as an example to introduce the active control system.

8.6.1 Modeling and Parameters

In this example, a three-story frame structure is selected. The structural parameters of the model are the mass vector $\{m\} = \{3, 3, 3\}^T \times 10^5$ kg and the initial stiffness vector $\{k\} = \{2, 2, 2\}^T \times 10^8$ N/m. The damping matrix of the structure adopts Rayleigh damping in which the first and the second damping ratios are both assumed as 0.05.

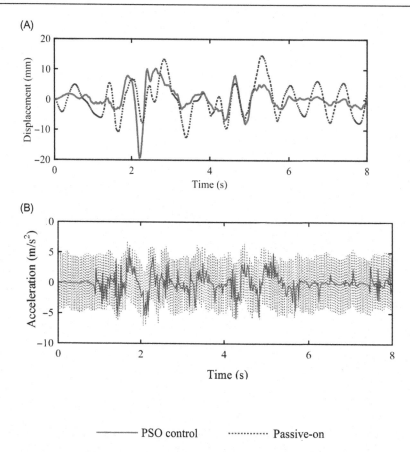

——— PSO control   ············ Passive-on

Figure 8.41
The top floor responses comparison between the PSO controlled structure and the passive-on controlled structure. (A) The displacement response and (B) the acceleration response.

El Centro earthquake with 200 gal acceleration amplitude is selected as excitation, and the sampling time is 0.02 s. The dynamic responses of structures with and without ATS can be calculated by MATLAB programming.

According to the Eq. (3.20), the motion equation of the ATS control system can be expressed as Eq. (2.4). The mass matrix, the stiffness matrix, and the damping matrix of model can be expressed as follows:

$$[M] = \begin{bmatrix} m_1 & 0 & 0 \\ 0 & m_2 & 0 \\ 0 & 0 & m_3 \end{bmatrix} = \begin{bmatrix} 3 & 0 & 0 \\ 0 & 3 & 0 \\ 0 & 0 & 3 \end{bmatrix} \times 10^5 \text{ kg},$$

$$[K] = \begin{bmatrix} k_1 + k_2 & -k_2 & 0 \\ -k_2 & k_2 + k_3 & -k_3 \\ 0 & -k_3 & k_3 \end{bmatrix} = \begin{bmatrix} 4 & -2 & 0 \\ -2 & 4 & -2 \\ 0 & -2 & 2 \end{bmatrix} \times 10^8 \text{ N/m},$$

and $[C] = \kappa_1[M] + \kappa_2[K]$.

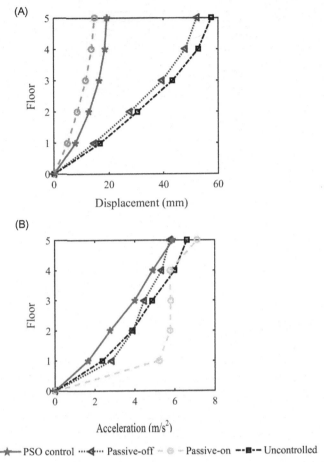

Figure 8.42
The maximum responses comparison of each floor under different control strategies. (A) The maximum displacement response and (B) the maximum acceleration response.

In this example, the active control devices are installed on each story of the structure, then the structure can be simplified, as shown in Fig. 8.43. Then the control force matrix and the position matrix can be expressed as

$$\{f_d(t)\} = \{f_{d1}, f_{d2}, f_{d3}\}^T \quad [B] = \begin{bmatrix} 1 & -1 & 0 \\ 0 & 1 & -1 \\ 0 & 0 & 1 \end{bmatrix} \quad (8.17)$$

As mentioned in Section 3.2, the state-space equation of the motion of ATS control system can be expressed as Eq. (2.5).

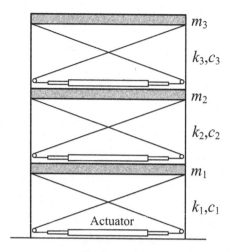

Figure 8.43
Active structure control system with ATS.

8.6.2 Active Control Strategy

In order to design an ATS, the active control force needs to be calculated firstly. LQR algorithm is the most widely used method in the design of the active control system, so the LQR optimal control algorithm is selected to design a state feedback controller $f_d(t)$, which makes the quadratic cost function minimum. Based on the explanation of the LQR algorithm in Section 2.2, this section mainly describes the steps to calculate the active control force by LQR algorithm.

Weight matrices $[Q]$ and $[R]$ are two vital control parameters in the LQR control algorithm, which determine the values of the control force and the structure response. The parameters $[Q]$ and $[R]$ can be expressed as:

$$[Q] = \alpha \begin{bmatrix} [K] & 0 \\ 0 & [M] \end{bmatrix} \quad [R] = \beta[I] \tag{8.18}$$

where α and β are undetermined coefficients, $[I]$ is a 3×3 unit matrix.

According to Section 3.2.3, the greater the $[Q]$ is and the smaller the $[R]$ is, then the smaller the structural responses and the greater of driving forces are. So in order to achieve better effect on active control, α is assumed as 100 and β is assumed as 1.0×10^{-7}.

The state feedback gain matrix can be gained by the function *lqr* in MATLAB, as follows:

$$[G] = lqr([A], [B_0], [Q], [R]) \tag{8.19}$$

Then, according to Eq. (3.21) in the Section 3.3.2, the optimal driving force should be $\{f_d(t)\} = -[G]\{Z(t)\}$, and the state-space equation can be rewritten as the Eq. (3.22). Finally, the above equation can be solved by the function *lsim* in MATLAB:

$$[y_0, \{Z\}] = lsim(([A] - [B_0][G]), [D], [C_1], [C_0], \{\ddot{x}_g\}, t) \tag{8.20}$$

where $[C_1]$ is the unit matrix, $[C_0]$ is the null matrix. The following program shows how to calculate the control force using the LQR algorithm in MATLAB:

```
B = [1 −1 0;0 1 −1;0 0 1];
Ds = −M*ones(3,1);
A = [zeros(3) eye(3);inv(−M)*K inv(−M)*C];
B0 = [zeros(3);inv(M)*Bs];
D0 = [zeros(3,1);inv(M)*Ds];
I = eye(3);
arph = 100;
beta = 1.0e − 7;
Q = arph*[K zeros(3);zeros(3) M];
R = beta*I;
G = lqr(A,B0,Q,R);
[Y,Z] = lsim((A − B0*G),D0,eye(6),zeros(6,1),xg,t);
```

8.6.3 Results and Analysis

The maximum inter-story displacement and the maximum acceleration response are calculated in MATLAB, as shown in Table 8.11.

It can be seen from Table 8.11 that the maximum inter-story displacement and the maximum acceleration can be reduced effectively by using active control. The maximum inter-story displacement of the first floor in the ATS controlled structure is 2.68 mm, by the reduction of 87.85%, compared with that of the uncontrolled structure, 22.07 mm. At the same time, the maximum acceleration response of the first floor in the ATS controlled structure is 1.29 m/s^2, by the reduction of 70.14%, compared with that of the uncontrolled structure, 4.32 m/s^2.

Table 8.11: The analysis results

Condition	The Maximum Inter-Story Displacement (mm)			The Maximum Acceleration (m/s^2)		
	1	2	3	1	2	3
Uncontrolled	22.07	16.82	9.07	4.32	6.47	7.42
ATS controlled	2.68	1.78	0.90	1.29	1.83	2.03

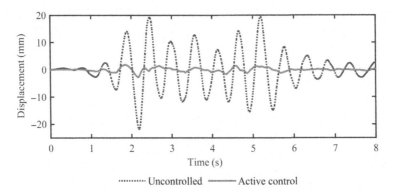

Figure 8.44
The displacement response of the first floor.

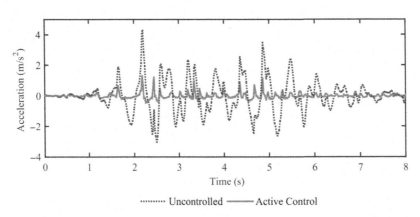

Figure 8.45
The acceleration response of the first floor.

The time-history curve of the displacement response of the first floor is shown in Fig. 8.44. Similarly, the time-history curve of the acceleration responses of the first floor is shown in Fig. 8.45.

It can be seen that the displacement and the acceleration responses of the structure are reduced obviously by using the ATS control system. This also shows that the active control method has obviously vibration-suppressing effect when civil engineering structures are subjected to earthquake or strong wind excitations. However, the active control involves in many complex techniques, such as the strong power requirement, controller design, time-delay effect, therefore promotion and a lot of real applications of active control still require a period of time.

References

[1] Causevic M, Mitrovic S. Comparison between non-linear dynamic and static seismic analysis of structures according to European and US provisions. Bull Earthquake Eng 2011;9(2):467–89.

[2] Cruz EF, Chopra AK. Simplified procedures for earthquake analysis of buildings. J Struct Eng 1986;112(3):461–80.

[3] Cruz EF, Chopra AK. Elastic earthquake response of building frames. J Struct Eng 1986;112(3):443–59.

[4] Newmark NM. A method of computation for structural dynamics. J Eng Mech Div 1959;85(3):67–94.

[5] Shinozuka M, Jan CM. Digital simulation of random processes and its applications. J Sound Vib 1972;25(1):111–28.

[6] Cebon D. Interaction between heavy vehicles and roads. Warrendale, PA, USA: SAE, Inc; 1993.

[7] Muto K. Seismic design of building. Tokyo: Maruzen Publishing Co., Ltd.; 1963.

[8] Passive and active structural vibration control in civil engineering. Springer; 2014.

[9] Xu ZD, Huang XH, Lu LH. Experimental study on horizontal performance of multi-dimensional earthquake isolation and mitigation devices for long-span reticulated structures. J Vib Control 2011; 1077546311418868.

[10] Xu ZD, Tu Q, Guo YF. Experimental study on vertical performance of multidimensional earthquake isolation and mitigation devices for long-span reticulated structures. J Vib Control 2012;18(13):1971–85.

[11] Nagarajaiah S, Feng MQ, Shinozuka M. Control of structures with friction controllable sliding isolation bearings. Soil Dyn Earthquake Eng 1993;12(2):103–12.

[12] Su L, Ahmadi G. Response of frictional base isolation systems to horizontal-vertical random earthquake excitations. Probab Eng Mech 1988;3(1):12–21.

[13] Liu S, Warn GP. Seismic performance and sensitivity of floor isolation systems in steel plate shear wall structures. Eng Struct 2012;42:115–26.

[14] Mostaghel N, Davis T. Representations of Coulomb friction for dynamic analysis. Earthquake Eng Struct Dyn 1997;26(5):541–8.

[15] Kelly JM, Hodder SB. Experimental study of lead and elastomeric dampers for base isolation systems. NASA STI/Recon Tech Rep N 1981;82:31577.

[16] Yeung N, Pan ADE. The effectiveness of viscous-damping walls for controlling wind vibrations in multi-story buildings. J Wind Eng Ind Aerod 1998;77:337–48.

[17] Housner GW, Bergman LA, Caughey TK, et al. Structural control: past, present, and future. J Eng Mech 1997;123(9):897–971.

[18] De Vicente J, Klingenberg DJ, Hidalgo-Alvarez R. Magnetorheological fluids: a review. Soft Matter 2011;7(8):3701–10.

[19] Brigadnov IA, Dorfmann A. Mathematical modeling of magnetorheological fluids. Continuum Mech Thermodyn 2005;17(1):29–42.

[20] Hagood NW, Chung WH, Von Flotow A. Modelling of piezoelectric actuator dynamics for active structural control. J Intell Mater Syst and Struct 1990;1(3):327–54.

[21] Struwe M. Variational methods: applications to nonlinear partial differential equations and Hamiltonian systems. Ergebnisse der Mathematik und ihrer Grenzgebiete. 3. Folge, Springer; 2008.

[22] Yang JN, Akbarpour A, Ghaemmaghami P. New optimal control algorithms for structural control. J Eng Mech 1987;113(9):1369–86.
[23] Yang JN, Akbarpour A, Ghaemmaghami P. Optimal control algorithms for earthquake-excited building structures. Structural Control. Netherlands:Springer.
[24] Welch G, Bishop G. An introduction to the Kalman filter, UNC-Chapel Hill, TR 95-041. Chapel Hill, NC: University of North Carolina; 2006.
[25] Catlin DE. Estimation, control, and the discrete Kalman filter. New York, USA: Springer Science & Business Media; 2012.
[26] Costa PJ. Adaptive model architecture and extended Kalman-Bucy filters. IEEE Trans Aerosp Electron Syst 1994;30(2):525–33.
[27] Ou JP. Structural vibration control: active, semi-active and intelligent controls. Beijing, China: Science Press; 2003.
[28] Yang JN. Application of optimal control theory to civil engineering structure. J Eng Mech Div 1975;101 (6):819–38.
[29] Chang JCH, Soong TT. Structural control using active tuned mass dampers. J Eng Mech Div 1980;106 (6):1091–8.
[30] Abdel-Rohman M, Leipholz HH. Active control of tall buildings. J Struct Eng 1983;109(3):628–45.
[31] Chung LL, Reinhorn AM, Soong TT. Experiments on active control of seismic structures. J Eng Mech 1988;114(2):241–56.
[32] Soong TT, Spencer Jr, Reviewer BF. Active structural control: theory and practice. J Eng Mech 1992;118 (6):1282–5.
[33] Suneja BP, Datta TK. Open-close loop active control of articulated leg platform. J Eng Mech 1998;124 (7):734–40.
[34] Porter B, Crossley R. Modal control: theory and applications. Abingdon, UK: Taylor & Francis Group; 1972.
[35] Abdel-Rohman M, Nayfeh AH. Active control of nonlinear oscillations in bridges. J Eng Mech 1987;113 (3):335–48.
[36] Abdel-Rohman M, Leipholz HHE. Structure control by pole assignment method. J Eng Mech Div, ASCE 1978;104:1157–75.
[37] Martin CR, Soong TT. Modal control of multistory structures. J Eng Mech Div 1976;102(4):613–23.
[38] Akbarpour A, Yang JN, Ghaemmaghami P. Instantaneous optimal control laws for tall buildings under seismic excitations. Technical Report NCEER-87-0007, June 10, 1987.
[39] Yang JN. Instantaneous optimal control for linear, nonlinear and hysteretic structures-stable controllers. Technical Report NCEER-91-0026, November 15, 1991.
[40] Yang JN, Li Z, Liu SC. Stable controllers for instantaneous optimal control. J Eng Mech 1992;118 (8):1612–30.
[41] Fang JQ, Li QS, Jeary AP. Modified independent modal space control of MDOF systems. J Sound Vib 2003;261(3):421–41.
[42] Soong TT. State-of-the-art review: active structural control in civil engineering. Eng Struct 1988;10 (2):74–84.
[43] Yang JN, Wu JC, Reinhorn AM, et al. Experimental verifications of $H\infty$ and sliding mode control for seismically excited buildings. J Struct Eng 1996;122(1):69–75.
[44] Yang JN, Wu JC, Agrawal AK. Sliding mode control for seismically excited linear structures. J Eng Mech 1995;121(12):1386–90.
[45] Yang JN, Wu JC, Agrawal AK, et al. Sliding mode control with compensator for wind and seismic response control. Earthquake Eng Struct Dyn 1997;26(11):1137–56.
[46] Housner GW, Soong TT, Masri SF. Second generation of active structural control in civil engineering. Comput-Aided Civ Infrastruct Eng 1996;11(5):289–96.
[47] Li J, Peng YB, Chen JB. A physical approach to structural stochastic optimal controls. Probab Eng Mech 2010;25(1):127–41.

[48] Peng YB, Li J. Exceedance probability criterion based stochastic optimal polynomial control of duffing oscillators. Int J Non Linear Mech 2011;46(2):457–69.

[49] Suhardjo J, Spencer BF, Sain MK. Non-linear optimal control of a duffing system. Int J Non Linear Mech 1992;27(2):157–72.

[50] Yang JN, Agrawal AK, Chen S. Optimal polynomial control for seismically excited nonlinear and hysteretic structures. Earthquake Eng Struct Dyn 1996;25(11):1211–30.

[51] Anderson BDO, Moore JB. Optimal control: linear quadratic methods. North Chelmsford, USA: Courier Corporation; 2007.

[52] Yang JN, Li Z, Vongchavalitkul S. A generalization of optimal control theory: linear and nonlinear structures. J Eng Mech 1994;120(2):266–83.

[53] Haykin S. Neural networks and learning machines. Upper Saddle River, NJ, USA: Pearson; 2009.

[54] Xu ZD, Shen YP, Guo YQ. Semi-active control of structures incorporated with magnetorheological dampers using neural networks. Smart Mater Struct 2003;12(1):80–7.

[55] Eberhart RC, Shi Y. Particle swarm optimization: developments, applications and resources 27–30 May. Proceedings of congress on evolutionary computation. Piscataway, NJ, Seoul, Korea: IEEE service center; 2001. p. 81–6.

[56] Kennedy J. The particle swarm: social adaptation of knowledge. Proceedings of international conference on evolutionary computation, Indianapolis, IN. Piscataway, NJ: IEEE Service Center; 1997.

[57] Reynolds CW. Flocks, herds and schools: a distributed behavioral model. Comput Graph 1987;21(4):25–34.

[58] Heppner F, Grenander U. A stochastic nonlinear model for coordinated bird flocks. In: Krasner S, editor. The ubiquity of chaos. Washington, DC: AAAS Publications; 1990.

[59] Wilson EO. Sociobiology: the new synthesis. Cambridge, MA: Belknap Press; 1975.

[60] Shi Y, Eberhart RC. A modified particle swarm optimizer. IEEE world congress on computational intelligence. Anchorage: IEEE; 1998. p. 69–73.

[61] Clerc M. The swarm and the queen: towards a deterministic and adaptive particle swarm optimization. Evolutionary computation, 1999. CEC 99. Proceedings of the 1999 congress on. IEEE. Washington, D.C. USA; 1999;3.

[62] Eberhart RC, Shi Y. Comparing inertia weights and constriction factors in particle swarm optimization. Evolutionary computation, 2000. Proceedings of the 2000 congress on. IEEE. California, USA; 2000;1:84–8.

[63] Riget J, Vesterstrøm JS. A diversity-guided particle swarm optimizer-the ARPSO. Department of Computer Science, University of Aarhus, Aarhus, Denmark. Tech Rep 2002; 2.

[64] Tang KS, Man KF, Kwong S, et al. Genetic algorithms and their applications. IEEE Signal Process Mag 1996;13(6):22–37.

[65] Koza JR. Genetic programming: a paradigm for genetically breeding populations of computer programs to solve problems. California, USA: Stanford University, Department of Computer Science; 1990.

[66] Holland JH. Adaptation in natural and artificial system: an introduction with application to biology, control and artificial intelligence. Ann Arbor: University of Michigan Press; 1975.

[67] Janikow CZ, Michalewicz Z. An experimental comparison of binary and floating point representations in genetic algorithms.. IEEE Comput Graph Appl 1991;31–6.

[68] Wright AH. Genetic algorithms for real parameter optimization. Found Genet Algorithms 1991;1:205–18.

[69] Golberg DE. Genetic algorithms in search, optimization, and machine learning. New York, USA: Addison Wesley; 1989.

[70] Cha YJ, Agrawal AK. Decentralized output feedback polynomial control of seismically excited structures using genetic algorithm. Struct Control Health Monit 2013;20(3):241–58.

[71] Jiang X, Adeli H. Neuro-genetic algorithm for non-linear active control of structures. Int J Numer Methods Eng 2008;75(7):770–86.

[72] Ricciardelli F, Pizzimenti AD, Mattei M. Passive and active mass damper control of the response of tall buildings to wind gustiness. Eng Struct 2003;25(9):1199–209.

[73] Tsai HC, Lin GC. Optimum tuned-mass dampers for minimizing steady-state response of support-excited and damped systems. Earthquake Eng Struct Dyn 1993;22(11):957−73.

[74] Aizawa S, Fukao Y, Minewaki S, Hayamizu Y, Abe H, Haniuda N. An experimental study on the active mass damper. In: Proceedings of 9th word conference on earthquake engineering, Tokyo-Kyoto, Japan, vol. 5; 1988. p. 871−6.

[75] Kobori T, Kanayama H, Kamagata S. Dynamic intelligent building as active seismic response controlled structure. In: Proceedings of annual meeting of the Architectural Institute of Japan, Tokyo, Japan; 1987.

[76] Soong TT, Reinhorn AM, Yang JN. Active response control of building structures under seismic excitation. In: Proceedings of 9th word conference on earthquake engineering, Tokyo-Kyoto, Japan, vol. 8;1988. p. 453−8.

[77] Aizawa S, Hayamizu Y, Higashino M, Soga Y, Yamamoto M. Experimental study of dual-axis active mass damper. Proceedings of 9th world conference on structural control research 1990;68−73.

[78] Li AQ, Qu WL, Cheng WR. Research on hybrid vibration control of Nanjing TV tower under wind excitation. J Build Struct 1996;17(3):9−16.

[79] Cao H, Reinhorn AM, Soong TT. Design of an active mass damper for a tall TV tower in Nanjing, China. Eng Struct 1997;20(3):134−43.

[80] Liu YH, Tan P, Zhou FL, Teng J, Yan WM. Study of dynamic performance of AMD control device driven by multiple linear motors in the canton tower. J Build Struct 2015;36(4):126−32.

[81] Soong TT. Active structural control: theory and practice. Essex, UK: Longman Scientific and Technical; 1990.

[82] Dyke SJ, Spencer Jr BF, Quast P, et al. Experimental verification of acceleration feedback control strategies for an active tendon system. Nat Center for Earthquake Engrg Res, Tech Report NCEER-94, 1994; 24.

[83] Yang JN, Samali B. Control of tall buildings in along-wind motion. J Struct Eng 1983;109(1):50−68.

[84] Roorda J. Experiments in feedback control of structures. Struct Control 1980;629−61.

[85] Bossens F, Preumont A. Active tendon control of cable-stayed bridges: a large-scale demonstration. Earthquake Eng Struct Dyn 2001;30(7):961−79.

[86] Nudehi S, Mukherjee R, Shaw SW. Active vibration control of a flexible beam using a buckling-type end force. J Dyn Syst Meas Control 2006;128(2):278−86.

[87] Issa J, Mukherjee R, Shaw SW. Vibration suppression in structures using cable actuators. J Vib Acoust 2010;132(3):031006.

[88] Reinhorn AM, Soong TT, Riley MA, et al. Full-scale implementation of active control. II: installation and performance. J Struct Eng 1993;119(6):1935−60.

[89] Miller RK, Masri SF, Dehghanyar TJ, et al. Active vibration control of large civil structures. J Eng Mech 1988;114(9):1542−70.

[90] Klein RE, Cusano C, Stukel JJ. Investigation of a method to stabilize wind-induced oscillations in large structures. Mechanical engineering. 345 E 47th St, New York, NY 10017: ASME-AMER society of mechanical engineering 1973;95(2) 53-53.

[91] Gupta H, Soong TT, Dargush GF. Active aerodynamic bidirectional control of structures I: modeling and experiments. Eng Struct 2000;22(4):379−88.

[92] Masri SF, Bekey GA, Caughey TK. On-line control of nonlinear flexible structures. J Appl Mech 1982;49 (4):877−84.

[93] Fujino Y, Warnitchai P, Pacheco BM. Active stiffness control of cable vibration.. J Appl Mech 1993;60 (4):948−53.

[94] Traina MI, Masri SF, Miller RK. An experimental study of the earthquake response of building models provided with active damping devices. Proceedings 9th world conference on earthquake engineering 1988;8:447−52.

[95] Kobayashi H, Nagaoka H. Active control of flutter of a suspension bridge. J Wind Eng Ind Aerod 1992;41(1):143−51.

[96] Nissen HD, Sørensen PH, Jannerup O. Active aerodynamic stabilisation of long suspension bridges. J Wind Eng Ind Aerod 2004;92(10):829−47.

[97] Karnopp D, Crosby MJ, Harwood RA. Vibration control using semi-active force generators. J Manuf Sci Eng 1974;96(2):619−26.
[98] Hrovat D, Barak P, Rabins M. Semi-active versus passive or active tuned mass dampers for structural control. J Eng Mech 1983;109(3):691−705.
[99] Symans MD, Constantinou MC. Semi-active control systems for seismic protection of structures: a state-of-the-art review. Eng Struct 1999;21(6):469−87.
[100] Rabinow J. The magnetic fluid clutch. Trans Am Inst Electr Eng 1948;2(67):1308−15.
[101] English JJF, Hornfeck AJ. Magnetic fluid clutch: U.S. Patent 2,650,684. 1953-9-1.
[102] Jacob R. Magnetic fluid torque and force transmitting device: U.S. Patent 2,575,360. 1951-11-20.
[103] Rosenfeld N, Wereley NM, Radakrishnan R, et al. Behavior of magnetorheological fluids utilizing nanopowder iron. Int J Mod Phys B 2002;16(17n18):2392−8.
[104] Goncalves FD, Carlson JD. An alternate operation mode for MR fluids—magnetic gradient pinch. Journal of physics: conference series. IOP Publishing 2009;149(1):012050.
[105] Wereley NM, Cho JU, Choi YT, et al. Magnetorheological dampers in shear mode. Smart Mater Struct 2008;17(1):015022.
[106] Dyke SJ, Spencer Jr BF, Sain MK, et al. Modeling and control of magnetorheological dampers for seismic response reduction. Smart Mater Struct 1996;5(5):565.
[107] Kamath GM, Hurt MK, Wereley NM. Analysis and testing of Bingham plastic behavior in semi-active electrorheological fluid dampers. Smart Mater Struct 1996;5(5):576−90.
[108] Mao M, Choi YT, Wereley NM. Effective design strategy for a magneto-rheological damper using a nonlinear flow model. Smart structures and materials. International society for optics and photonics 2005;446−55.
[109] Spencer Jr BF, Dyke SJ, Sain MK, et al. Phenomenological model for magnetorheological dampers. J Eng Mech 1997;123(3):230−8.
[110] Zhou Q. Two mechanic models for magneto-rheological damper and corresponding test verification. Earth Eng Eng Vib 2002;22(4):144−50 (in Chinese).
[111] Wereley NM, Pang L, Kamath GM. Idealized hysteresis modeling of electrorheological and magnetorheological dampers. J Intell Mater Syst and Struct 1998;9(8):642−9.
[112] Kamath GM, Wereley NM, Jolly MR. Characterization of magnetorheological helicopter lag dampers. J Am Helicopter Soc 1999;44(3):234−48.
[113] Bouc R. A mathematical model for hysteresis. Acta Acust United Acust 1971;24(1):16−25.
[114] Wen YK. Method for random vibration of hysteretic systems. J Eng Mech Div 1976;102(2):249−63.
[115] Dahl PR. Solid friction damping of mechanical vibrations.. Am Inst Aeronaut Astronaut J 1976;14 (12):1675−82.
[116] Xu ZD, Shen YP. Mathematical model and simulated analysis of MR dampers. J Build Struct 2003;33 (1):68−70 (in Chinese).
[117] Xu ZD, Jia DH, Zhang XC. Performance tests and mathematical model considering magnetic saturation for magnetorheological damper. J Intell Mater Syst and Struct 2012;23(12):1331−49.
[118] Spencer Jr BF, Yang CQ, Carlson JD, et al. Smart dampers for seismic protection of structures: a full-scale study. In: Proceedings of the second world conference on structural control, June 28−July 01, 1998.
[119] Ou JP, Guan XC. Experimental study of magnetorheological damper performance. Earthquake Eng Eng Vib 1999;19(4):76−81 (in Chinese).
[120] Li ZX, Wu LL, Xu LH, et al. Structural design of MR damper and experimental study for performance of damping force. Earthquake Eng Eng Vib 2003;23(1):128−32 (in Chinese).
[121] Moon SJ, Kim BH, Jeong JA. An experimental study on a magneto-rheological fluid damper for structural control subjected to base excitation. Trans Korean Soc Noise Vib Eng 2004;14:767−73.
[122] Qu WL, Liu J, Tu JW, Run M, Cheng HB. Crucial techniques for design of 500 kN large-scale MR damper.. Earthquake Eng Eng Vib 2007;2:124−30 (in Chinese).
[123] Deleted in Review.

[124] Li ZX, Lv Y, Xu LH, et al. Experimental studies on nonlinear seismic control of a steel-concrete hybrid structure using MR dampers. Eng Struct 2013;49:248–63.

[125] Li J, Mei Z, Chen J, et al. Experimental investigations of stochastic control of randomly base-excited structures. Adv Struct Eng 2012;15(11):1963–76.

[126] Sodeyama H, Sunakoda K, Suzuki K, et al. Development of large capacity semi-active vibration control device using magneto rheological fluid. ASME-Publications-PVP 2001;428:109–14.

[127] Chen ZQ, Wang XY, Ko JM, et al. MR damping system on Dongting Lake cable-stayed bridge. Smart structures and materials. International society for optics and photonics 2003;229–35.

[128] Fujitani H, Sodeyama H, Tomura T, et al. Development of 400kN magnetorheological damper for a real base-isolated building. In: Proceedings of SPIE-the international society for optical engineering, San Diego, America, March 2003 (5052). p. 265–76.

[129] Qu WL, Qi SQ, Tu JW, et al. Intelligent control for braking-induced longitudinal vibration responses of floating-type railway bridges. Smart Mater Struct. 2009;18(12):125003.

[130] Tu JW, Liu J, Qu WL, Cheng XD. Design and fabrication of 500-kN large-scale MR damper. J Intell Mater Syst and Struct 2011;22(5):475–87.

[131] Winslow WM. Method and means for translating electrical impulses into mechanical force: U.S. Patent 2,417,850. 1947-3-25.

[132] Wang L, Gong X, Wen W. Electrorheological fluid and its applications in microfluidics. Microfluidics. Berlin Heidelberg:Springer; 2011. p. 91–115.

[133] Wen W, Huang X, Yang S, et al. The giant electrorheological effect in suspensions of nanoparticles. Nat Mater 2003;2(11):727–30.

[134] Davis LC. Polarization forces and conductivity effects in electrorheological fluids. J Appl Phys 1992;72 (4):1334–40.

[135] Wen W, Huang X, Sheng P. Electrorheological fluids: structures and mechanisms. Soft Matter 2008;4 (2):200–10.

[136] Stanway R. Smart fluids: current and future developments. Mater Sci Technol 2004;20(8):931–9.

[137] Gamota DR, Filisko FE. Dynamic mechanical studies of electrorheological materials: moderate frequencies. J Rheol 1991;35(3):399–425. 1978-present.

[138] Pang L, Kamath GM, Wereley NM. Analysis and testing of a linear stroke magnetorheological damper. In: Proceedings of 39th AIAA/ASME/ASCE/AHS/ASC structures, structural dynamics, and materials conference and exhibit and AIAA/ASME/AHS adaptive structures forum; 1998. p. 2841–56.

[139] Hong SR, Choi SB, Choi YT, et al. A hydro-mechanical model for hysteretic damping force prediction of ER damper: experimental verification. J Sound Vib 2005;285(4):1180–8.

[140] Li B, Yang ZC. Experimental investigations on damping properties of electrorheological fluid damper. Chinese J Appl Mech 2003;20(1):81–4 (in Chinese).

[141] Liu P, Liu HJ, Teng J, Cao TY. Design and parameter identification of electro-rheological dampers. Funct Mater 2006;37(5):774–6 (in Chinese).

[142] Park YK, Choi SB. Vibration control of a cantilevered beam via hybridization of electro-rheological fluids and piezoelectric films. J Sound Vib 1999;225(2):391–8.

[143] Choi Y, Sprecher AF, Conrad H. Vibration characteristics of a composite beam containing an electrorheological fluid. J Intell Mater Syst and Struct 1990;1(1):91–104.

[144] Qu WL, Zhang YL, Guan JG, et al. Test and analysis of semiactive control of ER intelligent damper for structures. Earthquake Eng Eng Vib 1998;12(4):111–17 (in Chinese).

[145] Liu HJ, Yang ZC, Xi XM. Application of ER fluid in structural vibration control. Mech Sci Technol 1999;18(1):111–14 (in Chinese).

[146] Qu WL, Xu YL. Semi-active control for earthquake responses of reticulated shells with ER/MR smart dampers. Earthquake Eng Eng Vib 2001;21(4):24–31 (in Chinese).

[147] Hong SR, Choi SB, Han MS. Vibration control of a frame structure using electro-rheological fluid mounts. Mech Sci 2002;44:2027–45.

[148] Hong SR, Choi SB, Jung WJ, Jeong WB. Vibration isolation of structural systems using squeeze mode ER mounts. J Intell Mater Syst Struct 2002;13:421–4.

[149] Meitzler A, Tiersten HF, Warner AW, et al. IEEE standard on piezoelectricity. USA: IEEE; 1988.

[150] Chuanbing L, Changrong L, Yulin Z. Advances of research on piezo-intelligent structures. Piezoelectr Acoust 2002;24(1):42−7.
[151] Xu ZD, Xiang J, Wang XD. The smart piezoelectric friction dampers: CN ZL200810024640.0. 2008.
[152] Sun FP, Chaudhry ZA, Rogers CA, et al. Automated real-time structure health monitoring via signature pattern recognition. Smart structures & materials 95. International society for optics and photonics 1995;236−47.
[153] Zong ZH, Ren WX, Ruan Y. Recent advances in research on damage diagnosis for civil engineering structures. China Civ Eng J 2003;36(5):105−10 (in Chinese).
[154] Sun MQ, Li ZQ, Hou ZF. Application of piezoelectric materials in structural health monitoring of civil engineering structures. Concrete 2003;161(3):22−4 (in Chinese).
[155] Li HN, Zhao XY. Research and application of piezo-intelligent sensors in civil engineering. Earthquake Eng Eng Vib 2004;24(6):165−72 (in Chinese).
[156] Zhou Z, Ou JP. Comparative study of smart sensing material used in the smart monitoring of civil engineering. Archit Technol 2002;33(4):270−2.
[157] Qu J, Guan X, Wu B, et al. Smart piezoelectric-friction damper. Earthquake Eng Eng Vib 2000;20(1):81−6 (in Chinese).
[158] Ou JP, Yang Y. Piezoelectric-T shape variable friction damper and its performance tests and analysis. Earthquake Eng Eng Vib 2003;23(4):171−7 (in Chinese).
[159] Yang B, Ou JP, Liu GC. Performance tests and analysis of T shape PZT variable friction damper. Piezoelectr Acoust 2005;27(5):580−2 (in Chinese).
[160] Qu WL, Chen ZH, Xu YL. Wind-induced vibration control of high-rise steel-truss tower using piezoelectric smart friction damper. Earthquake Eng Eng Vib 2000;20(1):94−9 (in Chinese).
[161] Xu YL, Qu WL, Chen ZH. Control of wind-excited truss tower using semiactive friction damper. J Struct Eng 2001;127(8):861−8.
[162] Garrett GT, Chen G, Cheng FY, et al. Experimental characterization of piezoelectric friction dampers. SPIE's 8th annual international symposium on smart structures and materials. International society for optics and photonics 2001;405−15.
[163] Chen GD, Garrett GT, Chen CQ, et al. Piezoelectric friction dampers for earthquake mitigation of buildings: design, fabrication, and characterization. Struct Eng Mech 2004;17(3-4):539−56.
[164] Chen CQ, Chen GD. Shaking table tests of a quarter-scale three-storey building model with piezoelectric friction dampers. Struct Control Health Monit 2004;11(4):239−57.
[165] Li HN, Li J, Song GB. Improved seismic control of structure with variable friction dampers by GA. In: Proceedings of IEEE on intelligent control, Cyprus; 2005. p. 310−5.
[166] Li J, Li HN, Song GB. Semi-active vibration suppression using piezoelectric dampers based on bang-bang control laws. Third China-Japan-US symposium on structural control and health monitoring. China: Dalian; 2004.
[167] Song TT, Dargush GF. Passive energy dissipation systems in structural engineering. Chichester: Wiley; 1997.
[168] Dai NX, Tan P, Zou FL. Piezoelectric variable friction damper and its performance experiments and analysis. J Earthquake Eng Eng Vib 2013;33(3):205−14 (in Chinese).
[169] Dai NX. Experimental and theoretical research on the smart isolation system using PZT AND SMA complex friction dampers. Doctoral dissertation. Hunan: Hunan University; 2012 (in Chinese).
[170] Wang SL, Zhan M, Zhu XY, et al. Mechanical performance experiment of a new piezoelectric friction damper. Mater Rev B 2013;27(11):112−15 (in Chinese).
[171] Zhan M, Wang SL, Zhu JQ, et al. Vibration control tests of a model structure installed with piezoelectric friction damper with reset function. J Vib Shock 2015;34(14):45−50 (in Chinese).
[172] Li MX, Liu J. Semi-active control for structural vibration using variable stiffness with nonlinear damping. J Vib Eng 1998;11(3):333−9 (in Chinese).
[173] Huo LS, Li HN. Structural vibration control using semi-active variable stiffness tuned liquid column damper. J Vib Shock 2012;31(10):157−64 (in Chinese).
[174] Kori JG, Jangid RS. Semi-active stiffness dampers for seismic control of structures. Adv Struct Eng 2007;10(5):501−24.

[175] Gui LP, Li MX, Zhang YF, et al. Design and calculation of the electro-hydraulic variable stiffness device. Noise Vib Control 2000;4:15−17 (in Chinese).
[176] Li M, Liu J. Semiactive structural control with variable stiffness. J Vib Eng 1999;12(2):166−72 (in Chinese).
[177] Li M, Liu J. Experimental study of semiactive structural control using variable stiffness. Earthquake Eng Eng Vib 1998;18(4):90−6 (in Chinese).
[178] Kobori T. Experiment study on active variable stiffness system-active seismic response controlled structure. Proceedings of 4th world congress council on tall buildings and urban habitat. Hong Kong: The Hong Kong Polytechnic University; 1990. p. 561−72.
[179] Nasu T, Kobori T. Active variable stiffness system with non-resonant control. Earthquake Eng Struct Dyn 2001;30(11):1597−614.
[180] Kumar P, Jangid RS, Reddy GR. Response of piping system with semi-active variable stiffness damper under tri-directional seismic excitation. Nuc Eng Des 2013;258:130−43.
[181] Nasu T, Kobori T, Takahashi M, et al. Analytical study on a high-rise building with the active variable stiffness system. Proceedings of the second world conference on structural control 1998;1:805−14.
[182] Jabbari F, Bobrow JE. Vibration suppression with resettable device. J Eng Mech 2002;128(9):916−24.
[183] Kurino H, Tagami J, Shimzu K, et al. Switching oil damper with built-in controller for structural control. J Struct Eng 2003;129(7):895−904.
[184] Nishitani A, Nitta Y, Ikeda Y. Semi-active structural-control based on variable slip-force level dampers. J Struct Eng 2003;129(7):933−40.
[185] Fukukita A, Saito T, Shiba K. Control effect for 20-story benchmark building using passive semiactive device. J Eng Mech 2004;130(4):430−6.
[186] Kurata N, Kobori T, Takahashi M, et al. Actual seismic response controlled building with semi-active damper system. Earthquake Eng Struct Dyn 1999;28(11):1427−47.
[187] Takahashi M, Kobori T, Nasu T, et al. Active response control of buildings for large earthquakes-seismic response control system with variable structural characteristics. Smart Mater Struct 1998;7(4):522−9.
[188] He WL, Agrawal AK, Mahmoud K. Control of seismically excited cable-stayed bridge using resetting semi-active stiffness dampers. J Bridge Eng 2001;6:376−84.
[189] Patten WN, Kuo CC, He Q, et al. Seismic structural control via hydraulic semi-active vibration dampers (SAVD). Proceedings of 1st world conference on structural control. 1994;83−9.
[190] Niwa N, Kobori T, Takahashi M, et al. Dynamic loading test and simulation analysis of full-scale semi-active hydraulic damper for structural control. Earthquake Eng Struct Dyn 2000;29(6):789−812.
[191] Kawashima K, Unjoh S, Iida H, et al. Effectiveness of the variable damper for reducing seismic response of highway bridges. Proceedings of second US−Japan workshop on earthquake protective systems for bridges. Japan: PWRI, Tsukuba Science City; 1992. p. 479.
[192] Kawashima K, Unjoh S. Variable dampers and variable stiffness for seismic control of bridges. Proceedings of international workshop on structural control 1993;283−97.
[193] Symans MD, Constantinou MC. Seismic testing of a building structure with a seismic-active fluid damper control system. Earthquake Eng Struct Dyn 1997;26(7):759−77.
[194] Gavin HP, Hanson RD, Filisko FE. Electrorheological dampers, part II: testing and Modelling. J Appl Mech 1996;63(3):676−82.
[195] Shinozuka M, Ghanem R. Use of variable dampers for earthquake protection of bridges. Proceedings of second US-Japan workshop on earthquake protective systems for bridges. Japan: Tsukuba Science City; 1992. p. 507−16.
[196] Patten WN. New life for the Walnut Creek Bridge via semi-active vibration control. Newsl Int Assoc Struct Control 1997;2(1):4−5.
[197] Gavin HP, Hanson RD, Filisko FE. Electrorheological dampers, part I: analysis and design. J Appl Mech 1996;63(3):669−75.
[198] Li H, Yuan XS, Wu B. Experimental study on structures with semi-active fluid dampers. J Vib Eng 2002;25(1):24−9 (in Chinese).
[199] Yang RL, Zhou XY, Yan WM, et al. Performance evaluation of semiactive structural control using variable dampers. J Vib Shock 2007;26(3):37−41 (in Chinese).

[200] Carlson JD, Jolly MR. MR fluid, foam and elastomer devices. Mechatronics 2000;10(4):555−69.
[201] Ginder JM, Clark SM, Schlotter WF, et al. Magnetostrictive phenomena in magnetorheological elastomers. Int J Mod Phys B 2002;16(17n18):2412−18.
[202] Lokander M, Reitberger T, Stenberg B. Oxidation of natural rubber-based magnetorheological elastomers. Polym Degrad Stab 2004;86(3):467−71.
[203] Chen L, Gong XL, Li WH. Microstructures and viscoelastic properties of anisotropic magnetorheological elastomers. Smart Mater Struct 2007;16(6):2645.
[204] Popp KM, Kröger M, Hua LW, et al. MRE Properties under shear and squeeze modes and applications. J Intell Mater Syst and Struct 2010;21(15):1471−7.
[205] Guan X, Dong X, Ou JP. Magnetostrictive effect of magnetorheological elastomer. J Magn Magn Mater 2008;320(3):158−63.
[206] Jolly MR, Carlson JD, Munoz BC. A model of the behaviour of magnetorheological materials. Smart Mater Struct 1996;5(5):607.
[207] Davis LC. Model of magnetorheological elastomers. J Appl Phys 1999;85(6):3348−51.
[208] Shen Y, Golnaraghi MF, Heppler GR. Experimental research and modeling of magnetorheological elastomers. J Intell Mater Syst and Struct 2004;15(1):27−35.
[209] Liao GJ, Gong XL, Kang CJ, et al. The design of an active−adaptive tuned vibration absorber based on magnetorheological elastomer and its vibration attenuation performance. Smart Mater Struct 2011;20 (7):075015.
[210] Du H, Li W, Zhang N. Semi-active variable stiffness vibration control of vehicle seat suspension using an MR elastomer isolator. Smart Mater Struct 2011;20(10):105003.
[211] Zhu JT. Research on vibration isolation and attenuation of the broadband excitation platform by using magnetorheological elastomers. Nanjing: Southeast University; 2013 (in Chinese).
[212] Wang YL. Preparation of rubber based iron particles composites and research of their application in the field of safety engineering as magnetorheological elastomers. Hefei: University of Science and Technology of China; 2006 (in Chinese).
[213] Chen L, Gong XL, Li WH. Effect of carbon black on the mechanical performances of magnetorheological elastomers. Polym Test 2008;27(3):340−5.
[214] Chen L, Gong XL, Jiang WQ, et al. Influence of plasticizer on the magnetorheological effect of magnetorheological elastomers. J Funct Mater 2006;37(5):703−5 (in Chinese).
[215] Zhang XZ. Study on the fabrication and mechanism of magnetorheological elastomers. Hefei: University of Science and Technology of China; 2005 (in Chinese).
[216] Li WH, Zhou Y, Tian TF. Viscoelastic properties of MR elastomers under harmonic loading. Rheologica Acta 2010;49(7):733−40.
[217] Zhu JT, Xu ZD, Guo YQ. Magnetoviscoelasticity parametric model of an MR elastomer vibration mitigation device. J Smart Mater Struct 2012;21(7):075034.
[218] Ginder JM, Nichols ME, Elie LD, et al. Controllable-stiffness components based on magnetorheological elastomers. SPIE's 7th annual international symposium on smart structures and materials. Newport Beach, CA, USA: International Society for Optics and Photonics; 2000. p. 418−25.
[219] Ginder JM, Schlotter WF, Nichols ME. Magnetorheological elastomers in tunable vibration absorbers.. SPIE's 8th annual international symposium on smart structures and materials. Newport Beach, CA, USA: International Society for Optics and Photonics; 2001. p. 103−10.
[220] Watson JR. Method and apparatus for varying the stiffness of a suspension bushing: U.S. Patent 5,609,353. 1997-3-11.
[221] Shiga T, Okada A, Kurauchi T. Magnetroviscoelastic behavior of composite gels. J Appl Polym Sci 1995;58(4):787−92.
[222] Jolly MR, Carlson JD, Muñoz BC, et al. The magnetoviscoelastic response of elastomer composites consisting of ferrous particles embedded in a polymer matrix. J Intell Mater Syst and Struct 1996;7 (6):613−22.
[223] Lokander M, Stenberg B. Improving the magnetorheological effect in isotropic magnetorheological rubber materials. Polymer Test 2003;22(6):677−80.

[224] Mysore P, Wang X, Gordaninejad F. Thick magnetorheological elastomers. SPIE smart structures and materials + nondestructive evaluation and health monitoring. San Diego, California, USA: International Society for Optics and Photonics; 2011: 797711-797711-13.

[225] Dong XM, Miao YU, Liao CR, et al. A new variable stiffness absorber based on magneto-rheological elastomer. Trans Nonferrous Met Soc China 2009;19:s611−15.

[226] Xu Z, Gong X, Liao G, et al. An active-damping-compensated magnetorheological elastomer adaptive tuned vibration absorber. J Intell Mater Syst and Struct 2010.

[227] Tu JW, Ren W, Wu P. Magnetic finite element analysis of the intelligent laminated MRE isolator. Noise Vib Control 2010;(5):169−72 (in Chinese).

[228] Zhu JT, Xu ZD, Guo YQ. Experimental and modeling study on magnetorheological elastomers with different matrices. J Mat Civil Eng 2013;25:1762−71.

[229] Xu ZD, Sha LF, Zhang XC, et al. Design, performance test and analysis on magnetorheological damper for earthquake mitigation. Struct Control Health Monit 2013;20(6):956−70.

[230] Xu ZD, Guo YQ. Fuzzy control method for earthquake mitigation structures with magnetorheological dampers. J Intell Mater Syst and Struct 2006;17(10):871−81.

[231] Jung HJ, Choi KM, Lee HJ. Design and application of MR damper-based control system with electromagnetic induction part for structural control 22−27 May, 2007 World forum on smart materials and smart structures technology (SMSST'07). China: CRC Press; 2008. p. 111.

[232] Lee HJ, Moon SJ, Jung HJ, et al. Integrated design method of MR damper and electromagnetic induction system for structural control. The 15th international symposium on: smart structures and materials & nondestructive evaluation and health monitoring. San Diego, California: International Society for Optics and Photonics; 2008: 69320S-69320S-10.

[233] Phillips RW. Engineering applications of fluids with a variable yield stress. Berkeley: University of California; 1969.

[234] Boozer AH. Ohm's law for mean magnetic fields. J Plasma Phys 1986;35(1):133−9.

[235] Nishimura I, Kobori T, Sakamoto M, et al. Active tuned mass damper. Smart Mater Struct 1992;1(4):306−11.

[236] Chang CC, Yang HTY. Control of buildings using active tuned mass dampers. J Eng Mech 1995;121(3):355−66.

[237] Liu J, Zhou YC, Lei LH. Analysis and optimum design of active structural control system (AMD). Earthquake Eng Eng Vib 1996;16(3):55−60.

[238] Lord Corporation Product Brochure & Technical Report; 2004.

[239] Liu Z. Controllable current amplifier for magnetorheological damper. Electr Eng 2007;33(7):28−32.

[240] Guo YQ, Liu T, Xu ZD, et al. Design and Experiment on Single-Chip Microprocessor for MRD coupling sensing and control. Int J Distrib Sens Networks 2012.

[241] MMA7260QT datasheet. <http://www.freescale.com/files/sensors/doc/data_sheet/MMA7260QT.pdf>.

[242] Barr M. Pulse width modulation. Embedded Systems Programming; 2001. p. 103−4.

[243] Atmega16 datasheet. <http://www.atmel.com/Images/doc2466.pdf >.

[244] Zhang XC, Xu ZD. Testing and modeling of a CLEMR damper and its application in structural vibration reduction. Nonlinear Dyn 2012;70(2):1575−88.

[245] Giberson MF. Two nonlinear beams with definitions of ductility. J Struct Div 1969.

[246] Zhou J, Wu XB. The calculation of tall and complex building structure. Beijing: China Electric Power Press; 2008 (in Chinese).

[247] Shinozuka M. Simulation of multivariate and multidimensional random process. J Acoust Soc Am 1971;49(1):357−67.

[248] Kaimal JC, et al. Spectral characteristics of surface layer turbulence. J R Meteorol Soc 1972;98:563−89.

[249] Xu ZD, Shen YP. Intelligent bi-state control for the structure with magnetorheological dampers. J Intell Mater Syst and Struct 2003;14(1):35−42.

[250] Xu ZD. Study on semi-active control of structures incorporated with magnetorheological dampers. Report of Postdoctoral Fellowship, Xi'an Jiaotong University, Xi'an, China. 2002 (in Chinese).

Index

Note: Page numbers followed by "*f*" and "*t*" refer to figures and tables, respectively.

A

Accurate value, fuzzification of, 47
Active aerodynamic wind deflector system, 17, 79
   analysis and tests, 82
   form and principles, 79–80
Active intelligent control system, 16–17, 16*f*, 63, 85
   active aerodynamic wind deflector system, 17
   active mass damper (AMD) system, 17
   active support system, 17
   active tendon control system, 17
   basic principles, 64
   buildup of systems, 63–64
   classification, 64–65
   design and parameters optimization on, 166–171
      fail-safe reliability, 171
      feedback gain, 167–169
      minimum energy principle, 169–171
   example, 243–248
      active control strategy, 246–247
      modeling and parameters, 243–245
      results and analysis, 247–248
   gas pulse generator system, 17
Active mass damper (AMD) system, 63, 65–74, 166, 171
   basic principles, 65–67
   construction and design, 67–69
   experiment and engineering example, 72–74
   mathematical models and structural analysis, 69–72
   of Nanjing TV tower, 72–74, 73*f*
Active support system, 17, 79
Active tendon system (ATS), 65, 74–79, 166, 243–244, 248
   basic principles, 74–75
   construction and design, 75–77
   experiment and engineering example, 77–79
Aerodynamic flap system, 81*f*
ANSOFT Maxwell, 162–163, 165
Anti-earthquake protection measures, 3–4
ATmega16, 175–176, 177*f*
   connection diagram of, 178*f*

B

Beam-column model, 183, 189*f*, 190–191
Bingham model, 94*f*, 227–230
   modified, 94–95, 95*f*
Boltzmann network, 50
Bouc–Wen hysteresis model, 96–97, 97*f*
Buckling-restrained brace (BRB), 13–14

C

Circum-Pacific seismic belt, 1–2
Closed-loop control system, 63, 64*f*
Coefficient weighted mean method, 49
Control algorithms, 21
   classical linear optimal control algorithm, 22–27
      linear quadratic gauss (LQG) optimal control, 25–27
      linear quadratic regulator (LQR) optimal control, 23–25
   fuzzy control, 45–49
      basic principle, 46
      fuzzy controller, design of, 46–49, 46*f*
      defuzzification, 49
      determination of basic domain, 46–47
      determination of rule base, 48
      fuzzification of accurate value, 47
      parameter selection, 47–48
      selection of membership function, 48
   genetic algorithm (GA), 56–61, 57*f*
      basic principle, 56–57
      control realization, 60–61
      procedure of, 57–60
   H_∞ feedback control, 37–38
   H_∞ norm, 36–37
   independent mode space control (IMSC), 33–35
      based on equation of motion, 34–35
      based on state space, 33–34
   instantaneous optimal control algorithm, 30–32
   neural network control, 50–52
      basic principle, 50
      learning method, 51–52

Control algorithms (*Continued*)
  optimal polynomial control, 41–45
    applications, 45
    basic principle, 42–44
  particle swarm optimization control, 52–56
    basic principle, 53–55
    design procedure of the PSO algorithm, 55–56
  pole assignment method, 27–30, 30*f*
    with output feedback, 29–30
    with state feedback, 28
  sliding mode control (SMC), 38–41
    controller, design of, 40–41
    sliding surface, design of, 39–40
Controller, design of, 40–41

D

Dahl model and modified Dahl model, 97–98, 98*f*
Defuzzification, 46, 49
Demagnetization ability, good, 153
Design and parameters optimization on intelligent control devices, 151
  active control systems, 166–171
    fail-safe reliability, 171
    feedback gain, 167–169
    minimum energy principle, 169–171
  magnetorheological (MR) damper, 151–160
    design on, 151–156
    parameters optimization on, 156–160
  magnetorheological elastomers (MRE) device, 160–166
    magnetic circuit FEM simulation, 165–166
    parameter optimization for magnetic circuit, 160–165
Design of intelligent controller, 173–177
  acceleration responses collection, design of, 173–175

  microcontroller, design of, 175–177
  microcontroller chip, 176
  optical coupler, 177
  PWM technology, 175–176
Dielectric electrorheological (DER) effect, 111
Dongting Lake Bridge in Hunan, China, 108–109
Duration time, 3, 5
D-value method, 7
Dynamic response analysis, of intelligent control structure, 181
  elastic analysis, 181–187
    determination of control force of MR damper, 185–186
    mathematical model of structures, 181–184
    numerical analysis, 186–187
  elasto-plastic analysis method, 187–192
    elasto-plastic stiffness matrix, 190–192
    processing of turning points, 188–190, 189*f*
    restoring force model, 187–188
  SIMULINK, dynamic response analysis by, 192–199, 193*f*, 200*f*
    numerical analysis, 193–199
    simulation of the controlled structure, 192–193

E

Earthquake disaster, 1–5
Elastic analysis, 181–187
  determination of control force of MR damper, 185–186
  mathematical model of structures, 181–184
  numerical analysis, 186–187
Elasto-plastic analysis method, 187–192
  elasto-plastic stiffness matrix, 190–192
  processing of turning points, 188–190, 189*f*
  restoring force model, 187–188

Electrohydraulic servomechanism, 74
Electrorheological (ER) dampers, 18, 110–117
  analysis and design methods, 115
  basic principles, 110–112
  construction and design, 112–113
  dielectric electrorheological (DER) effect, 111
  ER behavior of ER fluid, 111*f*
  Kelvin chain element, 114
  mathematical models, 113–115
    postyield mechanisms, 114–115
    preyield mechanisms, 114
    yield force, 115
  operation modes, 112*f*
    flow mode, 112, 116*f*
    shear mode, 112
    squeeze-flow mode, 112
  tests and engineering applications, 115–117
  valve mode, 113*f*
Epicenter, 1
Equations of motion of intelligent control system, 21–22
Error function of the whole system, 51
Error rate quantization, 48
Eurasian seismic belt, 1–2
Excitation amplitude, 3
Experimental study, on intelligent controller, 177–180

F

Fail-safe reliability, 171
Feedback control, 63
  H_∞ feedback control, 37–38
Feedback gain
  design and parameters optimization based on, 167–169
Feedback network, 50
Feedforward control mode, 63
Feedforward network, 50
Finite element model, 162–163, 164*f*
Fluctuating wind, 6–7

Force−displacement hysteresis curve, 105, 137−138
Fourier spectrum, 3
Frame structure with MR dampers, dynamic analysis on, 201−207
   evaluation indicators of the control effect, 207t
   installation of MR dampers, 203f
   results and analysis, 204−207
   semiactive control strategy, 201−203, 203f
   structural and damper parameters, 201
Frequency spectrum, 3
Frequency-domain method, 3−4, 7
Friction dampers, 14
   piezoelectric, 117−126
Fuzzification
   of accurate value, 47
   of mathematical concepts, 46
Fuzzy control, 45−49
   basic principle, 46
Fuzzy controller, 230, 232−236
   accurate value, fuzzification of, 47
   basic domain, determination of, 46−47
   defuzzification, 49
   design of, 46−49, 46f
   membership function, selection of, 48
   parameter selection, 47−48
   rule base, determination of, 48
Fuzzy reasoning, 46, 49

G

Gas pulse generator system, 17
Genetic algorithm (GA), 56−61, 57f
   basic principle, 56−57
   control realization, 60−61
   procedure of, 57−60
      encoding scheme, 57
      fitness techniques, 58
      genetic operation, 59, 59f
      parent selection, 58
      replacement strategy, 60

Giberson single component principle, 190−191
Gravity method, 49
Gust pulsation. *See* Fluctuating wind

H

H_∞ feedback control, 37−38
H_∞ norm, 36−37
Hamiltonian function, 43−44
Hamilton−Jacobi theoretical framework, 42−43
Hamilton−Jacobi−Bellman equation, 43−44
Hopfield network, 50
Hybrid isolation system, 12
Hybrid mass damper (HMD) control strategy, 74
Hydrogen coolant explosion, 3

I

Independent mode space control (IMSC), 33−35
   based on equation of motion, 34−35
   based on state space, 33−34
Inertia force, formula of, 5
Intelligent control algorithm, 19−20
Interconnection network, 50
Inter-story isolation system, 12

J

Jacobi matrix, 52
Japan earthquake 2011, 3

K

Kalman filter, 25−27

L

Lagrange multiplier factor method, 23, 32
Levenberg−Marquart algorithm, 51−52
Linear quadratic gauss (LQG) optimal control, 25−27
Linear quadratic regulator (LQR) control algorithm, 23−25, 169−170, 246

   basic equation of, 23
   LQR optimal control algorithm, 211−213, 246
   solution of optimal control, 23−25
Long-span structure with MR dampers, dynamic analysis on, 208−215
   parameters and modeling, 208
   results and analysis, 213−215
   semiactive control strategy, 211−213
   wind load simulation, 209−211
LORD Company, ER dampers of, 101, 116−117
Lord SD-1005 MR dampers, 108−109
Love wave, 3
Lyapunov direct method, 40−41
Lyapunov function, 40, 43−44
Lyapunov matrices, 44

M

Maclaurin series, 42−43
Magnetic induction intensity, 152, 156, 163−165, 164f, 165t, 168t
Magnetic permeability, high, 152
Magnetic saturation mathematical model, 99, 99f
Magnetic-induced modulus, 138−139, 142, 160
Magnetorheological (MR) dampers, 18, 87−110, 102f, 151−160, 173, 181
   analysis and design methods, 99−100
   assembly diagram of, 160f
   basic principles, 87−89, 88f
      direct-shear mode, 89, 91f
      magnetic gradient pinch mode, 89
      spring mechanisms, 94f
      squeeze mode, 89
      valve mode, 88, 90f, 92f
   construction and design, 89−93
   coupling sensing and control, 179f
   design on, 151−156
      design principle, 154−155

Magnetorheological (MR) dampers (*Continued*)
  geometry design, 155
  magnetic circuit design, 156
  materials selection, 151–154
  frame structure with, dynamic analysis on, 201–207
    results and analysis, 204–207
    semiactive control strategy, 201–203
    structural and damper parameters, 201
  geometric parameters of, 102t
  installation on Dongting Lake Bridge, Hunan, China, 108–109, 109f
  installation on Xi'erhuan Bridge, Hanzhong, China, 109–110, 109f
  long-span structure with, dynamic analysis on, 208–215
    parameters and modeling, 208
    results and analysis, 213–215
    semiactive control strategy, 211–213
    wind load simulation, 209–211
  mathematical models, 94–99
    Bingham model and modified Bingham model, 94–95, 94f, 95f
    Bouc–Wen hysteresis model, 96–97, 97f
    Dahl model and modified Dahl model, 97–98, 98f
    magnetic saturation mathematical model, 99, 99f
    nonlinear hysteretic biviscous model, 95–96, 96f
    Sigmoid model, 98–99, 98f
  parameters optimization on, 156–160
    geometric optimization, 156–158
    magnetic circuit optimization, 158–160
  performance test on, 103f
  rheological mechanism of MR fluid, 87f
  structure incorporated with, 182f
  tests and engineering applications, 101–110
Magnetorheological (MR) intelligent controller, 174f
Magnetorheological elastomer (MRE) device, 19, 138–149, 160–166
  analysis and design methods, 144–145
    damping device, 145
    vibration absorber, 144–145
  basic principles, 138–140
  construction and design, 140–142
  macro parameter model for, 143f
  magnetic circuit FEM simulation, 165–166
  magnetic circuit structure, 142, 161f
  mathematical models, 142–144
  operation modes, 140f
  parameter optimization for magnetic circuit, 160–165
  performance test system, 147f
  platform with, dynamic analysis on, 215–223
    modeling and parameters, 217–220
    results and analysis, 222–223
    semiactive control strategy, 220–221
  preparation, 141
  simplified magnetic circuit of, 162f
  tests and engineering applications, 145–149
  working principle, 144f
MATLAB, 201, 220, 247
Maximum membership degree method, 49
Mean wind, 6–7
Membership function, 46, 60
  selection of, 48
Metal dampers, 13–14
Microcontroller, design of, 175–177
  microcontroller chip, 176
  optical coupler, 177
  pulse-width modulation (PWM) technology, 175–176
Mild steel dampers, 13–14
Minimum energy principle
  design and parameters optimization based on, 169–171
MMA7260QT, 174–175, 174f
Modified Mercalli Intensity scale, 3
Multistate control strategy, 101f

N

Nanjing TV tower, AMD control system of, 72–74, 73f
Neural network control, 50–52
  basic principle, 50
  learning method, 51–52
Nonlinear hysteretic biviscous model, 95–96, 96f

O

Ocean ridge seismic belt, 1–2
Open-loop control system, 63, 64f
Optical coupler, 177
Optimal control algorithm, instantaneous, 30–32
Optimal control force vector, 34–35, 185
Optimal control forces, 63, 100, 127, 185–186, 201–202, 220–221
Optimal polynomial control, 41–45
  applications, 45
  basic principle, 42–44

P

Parallel-plate Bingham model, 154–155
Particle swarm optimization (PSO) control, 52–56, 240–241
  basic PSO algorithm, 53–54
  design procedure of PSO algorithm, 55–56
  example, 239–243
    PSO optimization control, 240–241
    results and analysis, 241–243

structural and damper parameters, 239–240
improved PSO algorithm, 54–55
Passive tuned dampers, 15–16
Performance index function (PIF), 22–23, 25
Piezoelectric actuator, 18, 118, 124–127
Piezoelectric friction damper, 18, 118–120, 119*f*, 122–124, 122*t*, 123*f*
Piezoelectricity friction dampers, 117–126
  analysis and design methods, 123–124
  basic principles, 118
  computing model of the piston, 120*f*
  construction and design, 118–119
  mathematical models, 119–122
  tests and engineering applications, 124–126
Plane bar element, 183–184, 183*f*
Platform with MRE devices, dynamic analysis on, 215–223
  acceleration response analysis, 224*f*
  modeling and parameters, 217–220
  parameters of the vibration control system, 219*t*
  results and analysis, 222–223
  semiactive control strategy, 220–221
  vibration isolation scheme, 217–218, 217*f*
Pole assignment method, 27–30, 30*f*
  with output feedback, 29–30
  with state feedback, 28
Polynomial controller, 44–45
Positive piezoelectric effect, 117
Power spectrum, 3, 222–223
Pulse generator system, 79
  analysis and tests, 81
  form and principles, 79
Pulse-width modulation (PWM) technology, 175–176
P-wave, 3

Q
Quadratic error function, 51
Quantization factors, 47–48
Quasi-static methods, 7–8

R
Rayleigh damping model, 184
Rayleigh wave, 3
Response spectrum method, 3, 5
Riccati and Lyapunov equations, 44
Riccati equation, 25–26, 32
Riccati matrix, 43–44
Riccati matrix equations, 38
Richter, Charles, 3
Richter magnitude, 3
Rubber bearing, 11–12
Rubber isolation, 11–12, 11*f*
Rule base, determination of, 48

S
San Fernando earthquake 1971, 5
Sandwich rubber isolation pad. *See* Rubber bearing
Sanwa Tekki Cooperation, 101
Seismic action, 3–5
Seismic belts around the world, 2*f*
Seismic intensity, 3
Seismic responses, 3–4, 7, 226–227, 232–233
Seismic structure, 2*f*
Seismic waves, 3
Self-organizing network, 50
Semiactive control variable stiffness damper (SAVSD), 131
Semiactive intelligent control, 18–19, 85, 86*f*
  basic principles, 85–86
  classification, 86
  electrorheological (ER) dampers, 18, 110–117
    analysis and design methods, 115
    basic principles, 110–112
    construction and design, 112–113
    mathematical models, 113–115
    tests and engineering applications, 115–117
  magnetorheological (MR) dampers, 18, 87–110
    analysis and design methods, 99–100
    basic principles, 87–89
    construction and design, 89–93
    mathematical models, 94–99
    tests and engineering applications, 101–110
  magnetorheological elastomer (MRE) device, 19, 138–149
    analysis and design methods, 144–145
    basic principles, 138–140
    construction and design, 140–142
    mathematical models, 142–144
    tests and engineering applications, 145–149
  piezoelectricity friction dampers, 18, 117–126
    analysis and design methods, 123–124
    basic principles, 118
    construction and design, 118–119
    mathematical models, 119–122
    tests and engineering applications, 124–126
  semi-active TLD, 19
  semi-active TMD, 19
  semiactive varied damping damper, 19, 132–138
    analysis and design methods, 136–137
    basic principles, 133
    construction and design, 133–134
    mathematical model, 134–136
    tests and engineering applications, 137–138
  semiactive varied stiffness damper, 18–19, 126–132

Semiactive intelligent control (*Continued*)
   analysis and design methods, 129–131
   basic principles, 126–127
   construction and design, 127
   mathematical models, 128–129
   tests and engineering applications, 131–132
   shape memory alloy (SMA) dampers, 19
SFH618A, 177
   connection diagram of, 178f
Shape coefficient of wind load, 7
Shape memory alloy (SMA) dampers, 19
Short columns isolation system, 12
Sigmoid model, 98–99, 98f
SIMULINK, dynamic response analysis by, 192–199, 193f, 200f
   numerical analysis, 193–199
   simulation of controlled structure, 192–193
SIMULINK analysis, 223–239
   controlled structure, example of, 226–239
   structure without dampers, example of, 224–226
SIMULINK model of MR damper, 234f
Single degree of freedom (SDOF) elastic systems, 5
Sliding isolation, 12
Sliding mode control (SMC), 38–41
   controller, design of, 40–41
   sliding surface, design of, 39–40
Slotted Bolted Connection (SBC) friction damper, 124–126
Stable wind. *See* Mean wind
Static analysis method, 5
Story shear model, 181–183, 182f
Structure vibration control technology, 8–20
   basic principles, 8–9
   intelligent control, 16–20
      active. *See* Active intelligent control system
      intelligent control algorithm, 19–20
      semiactive. *See* Semiactive intelligent control
   vibration isolation, 9–12
      hybrid isolation, 12
      inter-story isolation, 12
      rubber isolation, 11–12, 11f
      short columns isolation, 12
      sliding isolation, 12
      suspension isolation, 12
   vibration mitigation, 13–16, 13f
      friction dampers, 14
      metal dampers, 13–14
      passive tuned dampers, 15
      viscoelastic (VE) dampers, 15
      viscous dampers, 14–15
Structure vibration isolation, 11f
Suspension isolation structures, 12
S-wave, 3

T

Tangshan earthquake, 3, 4f
Threefold line stiffness retrograde model, 187, 188f
Time history analysis method, 5, 186
Time-domain method, 3–4, 7
Time-history method, 192
Triangle shape dampers, 13–14
Tuned liquid damper (TLD), 15–16
Tuned mass damper (TMD), 15, 65–68, 167–169, 171
Turning points, processing of, 188–190, 189f

V

Variable damping damper, semiactive, 19
Variable stiffness damper, semiactive, 18–19
Variable stiffness device, 126–127, 128f
Varied stiffness system, 131–132
Vibration control, classification of, 10f
Vibration control mechanism, 8
Vibration isolation system, 9–12
   hybrid isolation, 12
   inter-story isolation, 12
   rubber isolation, 11–12, 11f
   short columns isolation, 12
   sliding isolation, 12
   suspension isolation, 12
Vibration mitigation system, 13–16, 13f
   friction dampers, 14
   metal dampers, 13–14
   passive tuned dampers, 15
   viscoelastic (VE) dampers, 15
   viscous dampers, 14–15
Viscoelastic (VE) dampers, 15
Viscoelastic–plastic model, 115
Viscous dampers, 14–15

W

Wenchuan earthquake, 3, 4f
Wilson-θ method, 186
Wind disaster, 6–8
Wind load
   dynamic simulations of, 7
   shape coefficient of, 7
Wind pressure, basic, 7
Wind speed time curve, 6–7
Wind velocity, 6
Wind vibration response analysis methods, 7

X

X-shape plate dampers, 13–14

Printed in the United States
By Bookmasters